SpringerBriefs in Electrical and Computer Engineering

More information about this series at http://www.springer.com/series/10059

Saleh Faruque

Radio Frequency Modulation Made Easy

Saleh Faruque
Department of Electrical Engineering
University of North Dakota
Grand Forks, ND
USA

ISSN 2191-8112   ISSN 2191-8120   (electronic)
SpringerBriefs in Electrical and Computer Engineering
ISBN 978-3-319-41200-9   ISBN 978-3-319-41202-3   (eBook)
DOI 10.1007/978-3-319-41202-3

Library of Congress Control Number: 2016945147

Printed on acid-free paper

This Springer imprint is published by Springer Nature
The registered company is Springer International Publishing AG Switzerland

Preface

By inventing the wireless transmitter or radio in 1897, the Italian physicist Tomaso Guglielmo Marconi added a new dimension to the world of communications. This enabled the transmission of the human voice through space without wires. For this epoch-making invention, this illustrious scientist was honored with the Nobel Prize for Physics in 1909. Even today, students of wireless or radio technology remember this distinguished physicist with reverence. A new era began in Radio Communications.

The classical Marconi radio used a modulation technique known today as "Amplitude Modulation" or just AM. This led to the development of Frequency Modulation (FM), amplitude shift keying (ASK), phase shift keying (PSK), etc. Today, these technologies are extensively used in various wireless communication systems. These modulation techniques form an integral part of academic curricula today.

This book presents a comprehensive overview of the various modulation techniques mentioned above. Numerous illustrations are used to bring students up-to-date in key concepts and underlying principles of various analog and digital modulation techniques. In particular, the following topics will be presented in this book:

- Amplitude Modulation (AM)
- Frequency Modulation (FM)
- Bandwidth occupancy in AM and FM
- Amplitude shift keying (ASK)
- Frequency shift keying (FSK)
- Phase shift keying (PSK)
- N-ary coding and M-ary modulation
- Bandwidth occupancy in ASK, FSK, and PSK

This text has been primarily designed for electrical engineering students in the area of telecommunications. However, engineers and designers working in the area of wireless communications would also find this text useful. It is assumed that the student is familiar with the general theory of telecommunications.

In closing, I would like to say a few words about how this book was conceived. It came out of my long industrial and academic career. During my teaching tenure at the University of North Dakota, I developed a number of graduate-level elective courses in the area of telecommunications that combine theory and practice. This book is a collection of my courseware and research activities in wireless communications.

I am grateful to UND and the School for the Blind, North Dakota, for affording me this opportunity. This book would never have seen the light of day had UND and the State of North Dakota not provided me with the technology to do so. My heartfelt salute goes out to the dedicated developers of these technologies, who have enabled me and others visually impaired to work comfortably.

I would like to thank my beloved wife, Yasmin, an English Literature buff and a writer herself, for being by my side throughout the writing of this book and for patiently proofreading it. My darling son, Shams, an electrical engineer himself, provided technical support in formulation and experimentation when I needed it. For this, he deserves my heartfelt thanks.

Finally, thanks are also to my doctoral student Md. Maruf Ahamed who found time in his busy schedule to assist me with the simulations, illustrations, and the verification of equations.

In spite of all this support, there may still be some errors in this book. I hope that my readers forgive me for them. I shall be amply rewarded if they still find this book useful.

Grand Forks, USA   Saleh Faruque
May 2016

Contents

1 Introduction to Modulation   1
   1.1 Background   1
   1.2 Modulation by Analog Signals   3
        1.2.1 AM, FM, and PM.   3
        1.2.2 AM and FM Bandwidth at a Glance   4
   1.3 Modulation by Digital Signal   5
        1.3.1 Amplitude Shift Keying (ASK) Modulation   5
        1.3.2 Frequency Shift Keying (FSK) Modulation.   6
        1.3.3 Phase Shift Keying (PSK) Modulation.   7
   1.4 Bandwidth Occupancy in Digital Modulation   7
        1.4.1 Spectral Response of the Encoded Data   8
        1.4.2 Spectral Response of the Carrier Frequency Before
               Modulation.   9
        1.4.3 ASK Bandwidth at a Glance. 10
        1.4.4 FSK Bandwidth at a Glance 11
        1.4.5 BPSK Bandwidth at a Glance. 12
   1.5 Conclusions 14
   References 15

2 Amplitude Modulation (AM) 17
   2.1 Introduction 17
   2.2 Amplitude Modulation 19
   2.3 AM Spectrum and Bandwidth. 20
        2.3.1 Spectral Response of the Input Modulating Signal. 20
        2.3.2 Spectral Response of the Carrier Frequency 21
        2.3.3 AM Spectrum and Bandwidth. 21
        2.3.4 AM Response Due to Low and High
               Modulating Signals 23
        2.3.5 AM Demodulation 24
        2.3.6 Drawbacks in AM. 24

2.4 Double Sideband-Suppressed Carrier (DSBSC)   25
    2.4.1 DSBSC Modulation.   25
    2.4.2 Generation of DSBSC Signal   26
    2.4.3 DSBSC Spectrum and Bandwidth   27
    2.4.4 DSBSC Drawbacks   28
2.5 Single Sideband (SSB) Modulation   29
    2.5.1 Why SSB Modulation?   29
    2.5.2 Generation of SSB-Modulated Signal.   29
    2.5.3 SSB Spectrum and Bandwidth   30
2.6 Conclusions   32
References   32

3 Frequency Modulation (FM)   33
3.1 Introduction   33
3.2 Frequency Modulation (FM).   34
    3.2.1 Background   34
    3.2.2 The Basic FM   35
3.3 FM Spectrum and Bandwidth   37
    3.3.1 Spectral Response of the Input Modulating Signal.   37
    3.3.2 Spectral Response of the Carrier Frequency   38
    3.3.3 FM Spectrum   39
    3.3.4 Carson's Rule and FM Bandwidth.   40
    3.3.5 Bessel Function and FM Bandwidth   41
    3.3.6 FM Bandwidth Dilemma   42
3.4 Conclusions   44
References   44

4 Amplitude Shift Keying (ASK)   45
4.1 Introduction   45
4.2 ASK Modulation.   46
4.3 ASK Demodulation   48
4.4 ASK Bandwidth   49
    4.4.1 Spectral Response of the Encoded Data   49
    4.4.2 Spectral Response of the Carrier Frequency Before
           Modulation.   51
    4.4.3 ASK Bandwidth at a Glance.   51
4.5 BER Performance   53
4.6 Conclusions   54
References   55

5 Frequency Shift Keying (FSK)   57
5.1 Introduction   57
5.2 Frequency Shift Keying (FSK) Modulation.   58
5.3 Frequency Shift Keying (FSK) Demodulation   60

5.4 FSK Bandwidth. 61
  5.4.1 Spectral Response of the Encoded Data 61
  5.4.2 Spectral Response of the Carrier Frequency Before
         Modulation. 63
  5.4.3 FSK Bandwidth at a Glance 63
5.5 BER Performance 65
5.6 Conclusions 67
References 67

6 **Phase Shift Keying (PSK)** 69
6.1 Introduction 69
6.2 Binary Phase Shift Keying (BPSK) 70
  6.2.1 BPSK Modulation. 70
  6.2.2 BPSK Demodulation 73
6.3 QPSK Modulation. 74
6.4 8PSK Modulation 75
6.5 16PSK Modulation 75
6.6 PSK Spectrum and Bandwidth 77
  6.6.1 Spectral Response of the Encoded Data 77
  6.6.2 Spectral Response of the Carrier Before Modulation 79
  6.6.3 BPSK Spectrum 79
6.7 Conclusions 82
References 82

7 **N-Ary Coded Modulation.** 85
7.1 Introduction 85
7.2 N-Ary Convolutional Coding and M-Ary Modulation 86
  7.2.1 Background 86
  7.2.2 Generation of Complementary Convolutional Codes 86
  7.2.3 2-Ary Convolutional Coding with QPSK Modulation 88
  7.2.4 4-Ary Convolutional Coding with 16PSK Modulation 89
7.3 N-Ary Convolutional Decoder. 91
  7.3.1 Correlation Receiver 91
  7.3.2 Error Correction Capabilities of N-Ary Convolutional
         Codes 93
7.4 N-Ary Orthogonal Coding and M-Ary Modulation 94
  7.4.1 Background 94
  7.4.2 Orthogonal Codes 95
  7.4.3 2-Ary Orthogonal Coding with QPSK Modulation 95
  7.4.4 4-Ary Orthogonal Coding with 16PSK Modulation 97
  7.4.5 2-Ary Orthogonal Decoding 97
  7.4.6 4-Ary Orthogonal Decoding 99
  7.4.7 Error Correction Capabilities of N-Ary Orthogonal
         Codes 99
7.5 Conclusions 103
References 104

Chapter 1
Introduction to Modulation

Topics

- Background
- Modulation by Analog Signal
- AM and FM Bandwidth at a Glance
- Modulation by Digital Signal
- ASK, FSK and PSK Bandwidth at a Glance

1.1 Background

Modulation is a technique that changes the characteristics of the carrier frequency in accordance with the input signal. Figure 1.1 shows the conceptual block diagram of a modern wireless communication system, where the modulation block is shown in the inset of the dotted block. As shown in the figure, modulation is performed at the transmit side and demodulation is performed at the receive side. This is the final stage of any radio communication system. The preceding two stages have been discussed elaborately in my previous book in this series [1, 2].

The output signal of the modulator, referred to as the modulated signal, is fed into the antenna for propagation. Antenna is a reciprocal device that transmits and receives the modulated carrier frequency. The size of the antenna depends on the wavelength (λ) of the sinusoidal wave where

$\lambda = c/f$ m
c = Velocity of light = 3×10^8 m/s
f = Frequency of the sinusoidal wave, also known as "carrier frequency."

Therefore, a carrier frequency much higher than the input signal is required to keep the size of the antenna at an acceptable limit. For these reasons, a high-frequency carrier signal is used in the modulation process. In this process, the

© The Author(s) 2017   1
S. Faruque, *Radio Frequency Modulation Made Easy*,
SpringerBriefs in Electrical and Computer Engineering,
DOI 10.1007/978-3-319-41202-3_1

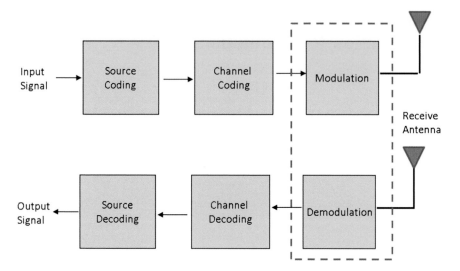

Fig. 1.1 Block diagram of a modern full-duplex communication system. The modulation stage is shown as a *dotted* block

low-frequency input signal changes the characteristics of the high-frequency carrier in a certain manner, depending on the modulation technique. Furthermore, as the size and speed of digital data networks continue to expand, bandwidth efficiency becomes increasingly important. This is especially true for broadband communication, where the digital signal processing is done keeping in mind the available bandwidth resources.

Hence, modulation is a very important step in the transmission of information. The information can be either analog or digital, where the carrier is a high-frequency sinusoidal waveform. As stated earlier, the input signal (analog or digital) changes the characteristics of the carrier waveform. Therefore, there are two basic modulation schemes as listed below:

- Modulation by analog signals and
- Modulation by digital signals.

This book presents a comprehensive overview of these modulation techniques in use today. Numerous illustrations are used to bring students up-to-date in key concepts and underlying principles of various analog and digital modulation techniques. For a head start, brief descriptions of each of these modulation techniques are presented below.

1.2   Modulation by Analog Signals

1.2.1   AM, FM, and PM

For analog signals, there are three well-known modulation techniques as listed below:

- Amplitude Modulation (AM),
- Frequency Modulation (FM),
- Phase Modulation (PM).

By inventing the wireless transmitter or radio in 1897, the Italian physicist Tomaso Guglielmo Marconi added a new dimension to the world of communications [3, 4]. This enabled the transmission of the human voice through space without wires. For this epoch-making invention, this illustrious scientist was honored with the Nobel Prize for Physics in 1909. Even today, students of wireless or radio technology remember this distinguished physicist with reverence. A new era began in Radio Communications. The classical Marconi radio used a modulation technique known today as "Amplitude Modulation" or just AM. In AM, the amplitude of the carrier changes in accordance with the input analog signal, while the frequency of the carrier remains the same. This is shown in Fig. 1.2 where

- $m(t)$ is the input modulating audio signal,
- $C(t)$ is the carrier frequency, and
- $S(t)$ is the AM-modulated carrier frequency.

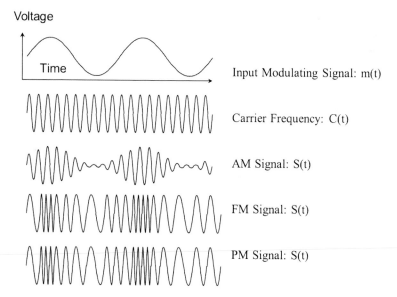

Fig. 1.2 Modulation by analog signal

As shown in the figure, the audio waveform changes the amplitude of the carrier to determine the envelope of the modulated carrier. This enables the receiver to extract the audio signal by demodulation. Notice that the amplitude of the carrier changes in accordance with the input signal, while the frequency of the carrier does not change after modulation. It can be shown that the modulated carrier $S(t)$ contains several spectral components, requiring frequency-domain analysis, which will be addressed in Chap. 2. It may be noted that AM is vulnerable to signal amplitude fading.

In Frequency Modulation (FM), the frequency of the carrier changes in accordance with the input modulation signal as shown in Fig. 1.2 [5]. Notice that in FM, only the frequency changes while the amplitude remains the same. Unlike AM, FM is more robust against signal amplitude fading. For this reason, FM is more attractive in commercial FM radio. In Chap. 3, it will be shown that in FM, the modulated carrier contains an infinite number of sideband due to modulation. For this reason, FM is also bandwidth inefficient.

Similarly, in Phase Modulation (PM), the phase of the carrier changes in accordance with the phase of the carrier, while the amplitude of the carrier does not change. PM is closely related to FM. In fact, FM is derived from the rate of change of phase of the carrier frequency. Both FM and PM belong to the same mathematical family. We will discuss this more elaborately in Chap. 3.

1.2.2 AM and FM Bandwidth at a Glance

The bandwidth occupied by the modulated signal depends on bandwidth of the input signal and the modulation method as shown in Fig. 1.3. Note that the unmodulated carrier itself has zero bandwidth.

In AM:

- The modulated carrier has two sidebands (upper and lower) and
- Total bandwidth = 2 × base band.

In FM:

- The carrier frequency shifts back and forth from the nominal frequency by Δf, where Δf is the frequency deviation.
- During this process, the modulated carrier creates an infinite number of spectral components, where higher-order spectral components are negligible.
- The approximate FM bandwidth is given by the Carson's rule:

  - FM BW = $2f(1 + \beta)$
  - f = Base band frequency
  - β = Modulation index
  - $\beta = \Delta f / f$
  - Δf = Frequency deviation.

Fig. 1.3 Bandwidth
occupancy in AM, FM, and
PM signals

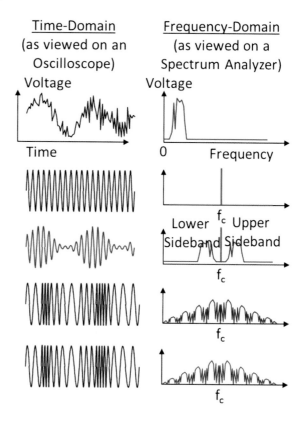

1.3 Modulation by Digital Signal

For digital signals, there are several modulation techniques available. The three main digital modulation techniques are as follows:

- Amplitude shift keying (ASK),
- Frequency shift keying (FSK), and
- Phase shift keying (PSK).

Figure 1.4 illustrates the modulated waveforms for an input modulating digital signal. Brief descriptions of each of these digital modulation techniques along with the respective spectral responses and bandwidth are presented below.

1.3.1 Amplitude Shift Keying (ASK) Modulation

Amplitude shift keying (ASK), also known as on–off keying (OOK), is a method of digital modulation that utilizes amplitude shifting of the relative amplitude of the

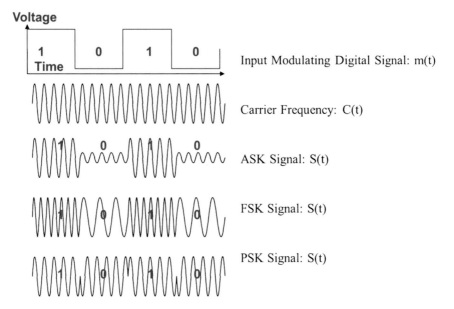

Fig. 1.4 Modulation by digital signal

career frequency [6–8]. The signal to be modulated and transmitted is binary; this is referred to as ASK, where the amplitude of the carrier changes in discrete levels, in accordance with the input signal, as shown.

- Binary 0 (bit 0): Amplitude = Low and
- Binary 1 (bit 1): Amplitude = High.

  Figure 1.4 shows the ASK-modulated waveform where

- Input digital signal is the information we want to transmit.
- Carrier is the radio frequency without modulation.
- Output is the ASK-modulated carrier, which has two amplitudes corresponding to the binary input signal. For binary signal 1, the carrier is ON. For the binary signal 0, the carrier is OFF. However, a small residual signal may remain due to noise, interference, etc.

1.3.2 *Frequency Shift Keying (FSK) Modulation*

Frequency shift keying (FSK) is a method of digital modulation that utilizes frequency shifting of the relative frequency content of the signal [6–8]. The signal to be modulated and transmitted is binary; this is referred to as binary FSK (BFSK), where the carrier frequency changes in discrete levels, in accordance with the input signal as shown below:

- Binary 0 (bit 0): Frequency $= f + \Delta f$.
- Binary 1 (bit 1): Frequency $= f - \Delta f$.

Figure 1.4 shows the FSK-modulated waveform where

- Input digital signal is the information we want to transmit.
- Carrier is the radio frequency without modulation.
- Output is the FSK-modulated carrier, which has two frequencies ω_1 and ω_2, corresponding to the binary input signal.
- These frequencies correspond to the messages binary 0 and 1, respectively.

1.3.3 Phase Shift Keying (PSK) Modulation

Phase shift keying (PSK) is a method of digital modulation that utilizes phase of the carrier to represent digital signal [6–8]. The signal to be modulated and transmitted is binary; this is referred to as binary PSK (BPSK), where the phase of the carrier changes in discrete levels, in accordance with the input signal as shown below:

- Binary 0 (bit 0): $Phase_1 = 0°$.
- Binary 1 (bit 1): $Phase_2 = 180°$.

Figure 1.4 shows the modulated waveform where

- Input digital signal is the information we want to transmit.
- Carrier is the radio frequency without modulation.
- Output is the BPSK-modulated carrier, which has two phases φ_1 and φ_2 corresponding to the two information bits.

1.4 Bandwidth Occupancy in Digital Modulation

In wireless communications, the scarcity of RF spectrum is well known. For this reason, we have to be vigilant about using transmission bandwidth. The transmission bandwidth depends on the following:

- Spectral response of the encoded data,
- Spectral response of the carrier frequency, and
- Modulation type (ASK, FSK, PSK), etc.

Let us take a closer look!

1.4.1 Spectral Response of the Encoded Data

In digital communications, data is generally referred to as a non-periodic digital signal. It has two values:

- Binary-1 = High, Period = T.
- Binary-0 = Low, Period = T.

Also, data can be represented in two ways:

- Time-domain representation and
- Frequency-domain representation.

The time-domain representation (Fig. 1.5a), known as non-return-to-zero (NRZ), is given by

$$V(t) = V \qquad <0<t<T$$
$$= 0 \qquad \text{elsewhere} \tag{1.1}$$

The frequency-domain representation is given by "Fourier transform" [9]:

$$V(\omega) = \int_0^T V \cdot e^{-j\omega t} dt \tag{1.2}$$

$$|V(\omega)| = VT \left[\frac{\sin(\omega T/2)}{\omega T/2} \right]$$

$$P(\omega) = \left(\frac{1}{T} \right) |V(\omega)|^2 = V^2 T \left[\frac{\sin(\omega T/2)}{\omega T/2} \right]^2 \tag{1.3}$$

Here, $P(\omega)$ is the power spectral density. This is plotted in (Fig. 1.5b). The main lobe corresponds to the fundamental frequency and side lobes correspond to harmonic components. The bandwidth of the power spectrum is proportional to the frequency. In practice, the side lobes are filtered out since they are relatively insignificant with respect to the main lobe. Therefore, the one-sided bandwidth is given by the ratio $f/fb = 1$. In other words, the one-sided bandwidth $= f = f_b$, where $f_b = R_b = 1/T$, T being the bit duration.

The general equation for two-sided response is given by

$$V(\omega) = \int_{-\alpha}^{\alpha} V(t) \cdot e^{-j\omega t} dt$$

In this case, $V(\omega)$ is called two-sided spectrum of $V(t)$. This is due to both positive and negative frequencies used in the integral. The function can be a voltage

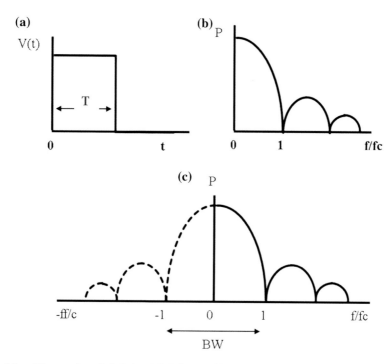

Fig. 1.5 a Discrete time digital signal, **b** it is one-sided power spectral density, and **c** two-sided power spectral density. The bandwidth associated with the non-return-to-zero (NRz) data is $2R_b$, where R_b is the bit rate

or a current (Fig. 1.5c) shows the two-sided response, where the bandwidth is determined by the main lobe as shown below:

$$\text{Two sided bandwidth (BW)} = 2R_b \ (R_b = \text{Bit rate before coding}) \quad (1.4)$$

1.4.2 Spectral Response of the Carrier Frequency Before Modulation

A carrier frequency is essentially a sinusoidal waveform, which is periodic and continuous with respect to time. It has one frequency component. For example, the sine wave is described by the following time-domain equation:

$$V(t) = V_p \sin(\omega t_c) \quad (1.5)$$

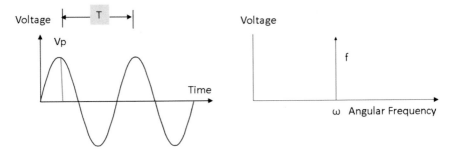

Fig. 1.6 A sine wave and its frequency response

where

$$V_p = \text{Peak voltage}$$

- $\omega_c = 2\pi f_c$
- f_c = Carrier frequency in Hz

Figure 1.6 shows the characteristics of a sine wave and its spectral response. Since the frequency is constant, its spectral response is located in the horizontal axis and the peak voltage is shown in the vertical axis. The corresponding bandwidth is zero.

1.4.3 ASK Bandwidth at a Glance

In ASK, the amplitude of the carriers changes in discrete levels, in accordance with the input signal where

- Input data: $m(t) = 0$ or 1.
- Carrier frequency: $C(t) = A_c \cos(\omega_c t)$.
- Modulated carrier: $S(t) = m(t)C(t) = m(t)A_c \cos(\omega_c t)$.

Since $m(t)$ is the input digital signal and it contains an infinite number of harmonically related sinusoidal waveforms and that we keep the fundamental and filter out the higher-order components, we write:

$$m(t) = A_m \sin(\omega_m t)$$

The ASK-modulated signal then becomes:

$$S(t) = m(t)S(t) = A_m A_c \sin(\omega_m t) \cos(\omega_c)$$
$$= A_m A_c \cos(\omega_c \pm \omega_m)$$

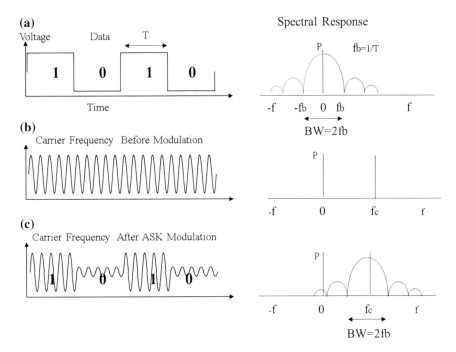

Fig. 1.7 ASK bandwidth at a glance. **a** Spectral response of NRZ data before modulation. **b** Spectral response of the carrier before modulation. **c** Spectral response of the carrier after modulation. The transmission bandwidth is $2f_b$, where f_b is the bit rate and $T = 1/f_b$ is the bit duration for NRZ data

The spectral response is depicted in Fig. 1.7. Notice that the spectral response after ASK modulation is the shifted version of the NRZ data. Bandwidth is given by, $BW = 2R_b$ (coded), where R_b is the coded bit rate.

1.4.4   FSK Bandwidth at a Glance

In FSK, the frequency of the carrier changes in two discrete levels, in accordance with the input signals. We have:

- Input data: $m(t) = 0$ or 1
- Carrier frequency: $C(t) = A \cos(\omega t)$
- Modulated carrier: $S(t) = A \cos(\omega - \Delta \omega)t$, For $m(t) = 1$
  $S(t) = A \cos(\omega + \Delta \omega)t$, For $m(t) = 0$

where

- $S(t)$ = The modulated carrier,
- A = Amplitude of the carrier,

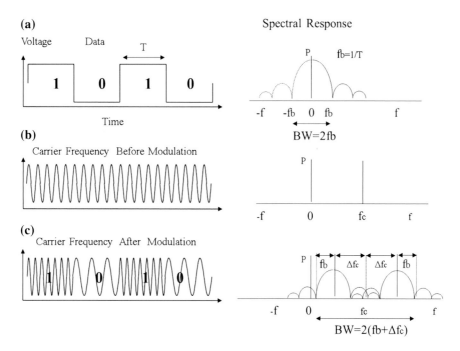

Fig. 1.8 FSK bandwidth at a glance. **a** Spectral response of NRZ data before modulation. **b** Spectral response of the carrier before modulation. **c** Spectral response of the carrier after modulation. The transmission bandwidth is $2(f_b + \Delta f_c)$. f_b is the bit rate and Δf_c is the frequency deviation $= 1/f_b$ is the bit duration for NRZ data

- ω = Nominal frequency of the carrier frequency, and
- $\Delta\omega$ = Frequency deviation.

The spectral response is depicted in Fig. 1.8. Notice that the carrier frequency after FSK modulation varies back and forth from the nominal frequency f_c by $\pm \Delta f_c$, where Δf_c is the frequency deviation. The FSK bandwidth is given by

$$\text{BW} = 2(f_b + \Delta f_c) = 2f_b(1 + \Delta f_c/f_b) = 2f_b(1 + \beta),$$

where $\beta = \Delta f/f_b$ is known as the modulation index and f_b is the coded bit frequency (bit rate R_b).

1.4.5 BPSK Bandwidth at a Glance

In BPSK, the phase of the carrier changes in two discrete levels, in accordance with the input signal. Here, we have:

- Input data: $m(t) = 0$ or 1
- Carrier frequency: $C(t) = A \cos (\omega t)$
- Modulated carrier: $S(t) = A \cos(\omega + \varphi)t$

where

- A = Amplitude of the carrier frequency,
- ω = Angular frequency of the carrier, and
- φ = Phase of the carrier frequency.

Table below shows the number of phases and the corresponding bits per phase for MPSK modulation schemes for $M = 2, 4, 8, 16, 32, 64$, etc. It will be shown that higher-order MPSK modulation schemes ($M > 2$) are spectrally efficient.

Modulation	Number of phases φ	Number of bits per phase
BPSK	2	1
QPSK	4	2
8PSK	8	3
16	16	4
32	32	5
64	64	6
:	:	:

Figure 1.9 shows the spectral response of the BPSK modulator. Since there are two phases, the carrier frequency changes in two discrete levels, one bit per phase, as follows:

$\varphi = 0°$ for bit 0 and
$\varphi = 180°$ for bit 1.

Notice that the spectral response after BPSK modulation is the shifted version of the NRZ data, centered on the carrier frequency f_c. The transmission bandwidth is given by

$$\text{BW(BPSK)} = 2R_b/\text{Bit per Phase} = 2R_b/1 = 2R_b$$

where

- R_b is the coded bit rate (bit frequency).
- For BPSK, $\varphi = 2$, one bit per phase.

Also, notice that the BPSK bandwidth is the same as the one in ASK modulation. This is due to the fact that the phase of the carrier changes in two discrete levels, while the frequency remains the same.

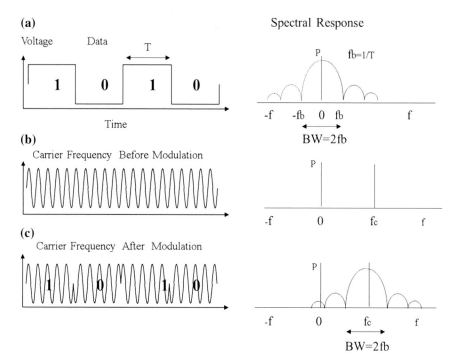

Fig. 1.9 PSK bandwidth at a glance. **a** Spectral response of NRZ data before modulation. **b** Spectral response of the carrier before modulation. **c** Spectral response of the carrier after modulation

1.5 Conclusions

This chapter presents a brief overview of modulation techniques covered in this book. Numerous illustrations are used to bring students up-to-date in key concepts and underlying principles of various analog and digital modulation techniques. In particular, following topics will be presented in this book:

- Amplitude Modulation (AM),
- Frequency Modulation (FM),
- Bandwidth occupancy in AM and FM,
- Amplitude shift keying (ASK),
- Frequency shift keying (FSK),
- Phase shift keying (PSK), and
- Bandwidth occupancy in ASK, FSK, and PSK.

References

1. Faruque S (2014) Radio frequency source coding made easy. Springer, New York
2. Faruque S (2014) Radio frequency channel coding made easy. Springer, New York
3. Marconi G (1987) Improvements in transmitting electrical impulses and signals, and in apparatus therefor. British patent No. 12,039 . Date of Application 2 June 1896; Complete Specification Left, 2 Mar 1897; Accepted, 2 July 1897 (later claimed by Oliver Lodge to contain his own ideas which he failed to patent)
4. Marconi G (1900) Improvements in apparatus for wireless telegraphy. British patent No. 7,777. Date of Application 26 Apr 1900; Complete Specification Left, 25 Feb 1901; Accepted, 13 Apr 1901
5. Armstrong EH (1936) A Method of reducing disturbances in radio signaling by a system of frequency modulation. Proc IRE (IRE) 24(5):689–740. doi:10.1109/JRPROC.1936.227383
6. Smith DR (1985) Digital transmission system. Van Nostrand Reinhold Co. ISBN: 0442009178
7. Leon W, Couch II (2001) Digital and analog communication systems, 7th edn. Prentice-Hall Inc, Englewood Cliffs. ISBN 0-13-142492-0
8. Sklar B (1988) Digital communications fundamentals and applications. Prentice Hall, Upper Saddle River
9. Joseph Fourier JB (1878) The analytical theory of heat (trans: Freeman A). The University Press, London

Chapter 2
Amplitude Modulation (AM)

Topics

- Introduction
- Amplitude Modulation (AM)
- AM Spectrum and Bandwidth
- Double Side Band Suppressed Carrier (DSBSC)
- DSBSC Spectrum and Bandwidth
- Single Side Band (SSB) Carrier
- SSB Spectrum and Bandwidth

2.1 Introduction

By inventing the wireless transmitter or radio in 1897, the Italian physicist Tomaso Guglielmo Marconi added a new dimension to the world of communications [1, 2]. This enabled the transmission of the human voice through space without wires. For this epoch-making invention, this illustrious scientist was honored with the Nobel Prize for Physics in 1909. Even today, students of wireless or radio technologies remember this distinguished physicist with reverence. A new era began in Radio Communications.

The classical Marconi radio used a modulation technique known today as "Amplitude Modulation" or just AM, which is the main topic of this chapter. In AM, the amplitude of the carrier changes in accordance with the input analog signal, while the frequency of the carrier remains the same. This is shown in Fig. 2.1, where

- $m(t)$ is the input modulating audio signal,
- $C(t)$ is the carrier frequency, and
- $S(t)$ is the AM-modulated carrier frequency.

© The Author(s) 2017
S. Faruque, *Radio Frequency Modulation Made Easy*,
SpringerBriefs in Electrical and Computer Engineering,
DOI 10.1007/978-3-319-41202-3_2

Fig. 2.1 AM waveforms.
The amplitude of the carrier
changes in accordance with
the input analog signal. The
frequency of the carrier
remains the same

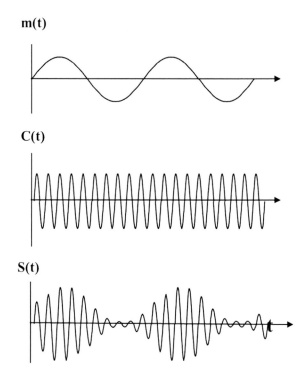

As shown in the figure, the audio waveform changes the amplitude of the carrier to determine the envelope of the modulated carrier. This enables the receiver to extract the audio signal by demodulation. Notice that the amplitude of the carrier changes in accordance with the input signal, while the frequency of the carrier does not change after modulation. However, it can be shown that the modulated carrier $S(t)$ contains several spectral components, requiring frequency domain analysis. In an effort to examine this, this chapter will present the following topics:

- Amplitude Modulation (AM) and AM spectrum
- Double Sideband-suppressed carrier (DSBSC) and DSBSC spectrum
- Single sideband (SSB) carrier and SSB spectrum

In the following sections, the above disciplines in AM modulation will be presented along with the respective spectrum and bandwidth. These materials have been augmented by diagrams and associated waveforms to make them easier for readers to grasp.

2.2 Amplitude Modulation

Amplitude Modulation (AM) is a method of analog modulation that utilizes amplitude variations of the relative amplitude of the career frequency [3–5]. The signal to be modulated and transmitted is analog. This is referred to as AM, where the amplitude of the carrier changes in accordance with the input signal.

Figure 2.2 shows a functional diagram of a typical AM modulator for a single tone. Here, $m(t)$ is the input analog signal we want to transmit, $C(t)$ is the carrier frequency without modulation, and $S(t)$ is the output AM-modulated carrier frequency. These parameters are described below:

$$
\begin{aligned}
\bullet \quad & m(t) = A_m \cos(2\pi f_m t) \\
\bullet \quad & C(t) = A_c \cos(2\pi f_c t) f_c \gg f_m \\
\bullet \quad & S(t) = [1 + m(t)]C(t) \\
& \quad\;\; = C(t) + m(t)C(t)
\end{aligned}
\tag{2.1}
$$

Therefore,
When $m(t) = 0$:

$$
\bullet \quad S(t) = A_c \cos(2\pi f_c t) \tag{2.2}
$$

When $m(t) = A_m \cos(2\pi f_m t)$:

$$
\bullet \quad S(t) = A_c \cos(2\pi f_c t) + A_c A_m \cos(2\pi f_m t) \cos(2\pi f_c t) \tag{2.3}
$$

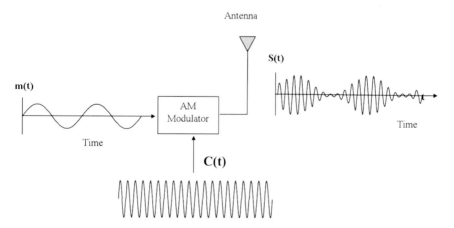

Fig. 2.2 Illustration of amplitude modulation. The amplitude of the carrier $C(t)$ changes in accordance with the input modulating signal $m(t)$. $S(t)$ is the modulated waveform which is transmitted by the antenna

In the above equation, we see that:

- The first term is the carrier only, which does not have information
- The second **term** contains **the** information, which has several spectral components, requiring further analysis to quantify them.

2.3   AM Spectrum and Bandwidth

In wireless communications, the scarcity of RF spectrum is well known. For this reason, we have to be vigilant about using transmission bandwidth and modulation. The transmission bandwidth depends on the following:

- Spectral response of the input modulating signal
- Spectral response of the carrier frequency and
- Modulation type (AM, FM ASK, FSK, PSK, etc.)

Let us take a closer look!

2.3.1   Spectral Response of the Input Modulating Signal

In AM, the input modulating signal is a continuous time low-frequency analog signal. For simplicity, we use a sinusoidal waveform, which is periodic and continuous with respect to time. It has one frequency component. For example, the sine wave is described by the following time domain equation:

$$V(t) = V_p \sin(\omega_m t) \tag{2.4}$$

where

- V_p = Peak voltage
- $\omega_m = 2\pi f_m$
- f_m = Input modulating frequency in Hz

Figure 2.3 shows the characteristics of a sine wave and its spectral response. Since the frequency is constant, its spectral response is located in the horizontal axis at f_m, and the peak voltage is shown in the vertical axis. The corresponding bandwidth is zero.

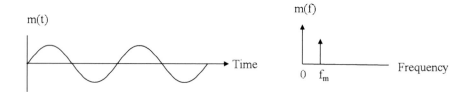

Fig. 2.3 A low-frequency sine wave and its frequency response

2.3.2 Spectral Response of the Carrier Frequency

A carrier frequency (f_c) is essentially a sinusoidal waveform, which is periodic and continuous with respect to time. It has one frequency component, which is much higher than the input modulating frequency $(f_c \gg f_m)$. For example, the sine wave is described by the following time domain equation:

$$V(t) = V_p \sin(\omega_c t) \tag{2.5}$$

where

$$V_p = \text{Peak voltage}$$

- $\omega_c = 2\pi f_c$
- $f_c =$ Carrier frequency in Hz

Figure 2.4 shows the characteristics of a sine wave and its spectral response. Since the frequency is constant, its spectral response is located in the horizontal axis at f_c and the peak voltage is shown in the vertical axis. The corresponding bandwidth is zero.

2.3.3 AM Spectrum and Bandwidth

Let us consider the AM signal again, which was derived earlier:

$$S(t) = A_c \cos(2\pi f_c t) + A_c A_m \cos(2\pi f_m t) \cos(2\pi f_c t) \tag{2.6}$$

Using the following trigonometric identity:

$$\cos A \cos B = 1/2\cos(A + B) + 1/2\cos(A - B) \tag{2.7}$$

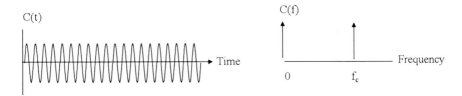

Fig. 2.4 A high-frequency sine wave and its frequency response

where

- $A = 2\pi (f_c + f_m)t$
- $B = 2\pi (f_c - f_m)t$

we get

$$S(t) = A_c \cos(2\pi f_c t) + (1/2)A_c A_m \cos[2\pi(f_c + f_m)t] + (1/2)A_c A_m \cos[2\pi(f_c - f_m)t]$$

$$(2.8)$$

This is the spectral response of the AM-modulated signal. It has three spectral components:

- The carrier: f_c
- Upper sideband: $f_c + f_m$
- Lower sideband: $f_c - f_m$

where f_c is the carrier frequency and f_m is the input modulating frequency. This is shown in Fig. 2.5. The AM bandwidth (BW) is given by

$$BW = 2f_m \qquad (2.9)$$

Notice that the power is distributed among the sidebands and the carrier, where the carrier does not contain any information. Only the sidebands contain the information. Therefore, AM is inefficient in power usage.

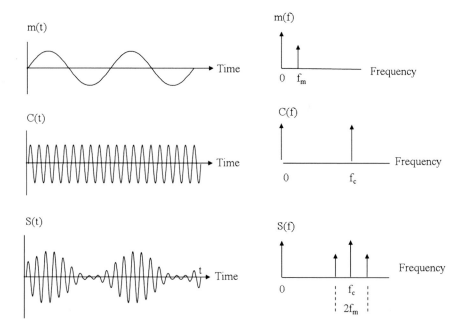

Fig. 2.5 AM spectrum. The bandwidth is given by $2f_m$

2.3.4   AM Response Due to Low and High Modulating Signals

If $m(t)$ has a peak positive value of less than +1 and a peak negative value of higher than −1, then the modulation is less than 100 %. This is shown in Fig. 2.7.

On the other hand, if $m(t)$ has a peak positive value of +1 and a peak negative value of −1, then the modulation is 100 % [3–5]. Therefore,

- For $m(t) = -1$: $S(t) = A_c[1 - 1]\cos(2\pi_c t) = 0$ (2.10)

- For $m(t) = +1$: $S(t) = A_c[1 + 1]\cos(2\pi f_c t)$
$= 2A_c \cos(2\pi f_c t)$ (2.11)

This is called 100 % modulation, as shown in Fig. 2.6. The percent modulation is described by the following equation:

The overall modulation percentage is:

$$\% \,\text{Overall Modulation} = \frac{A_{\max} - A_{\min}}{A_c} \times 100$$ (2.12)
$$= \frac{\max[m(t)] - \min[m(t)]}{2A_c} \times 100.$$

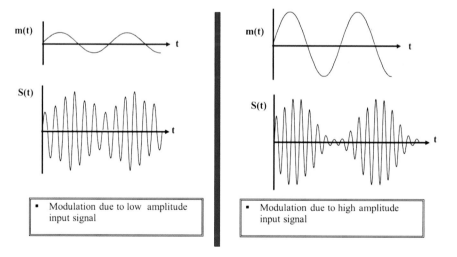

- Modulation due to low amplitude input signal

- Modulation due to high amplitude input signal

Fig. 2.6 Amplitude modulation due to low and high modulating signals

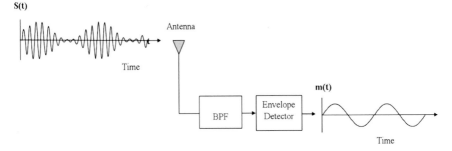

Fig. 2.7 AM demodulation technique. As the signal enters the receiver, it passes through the band-pass filter, which is tuned to the carrier frequency f_0. Next, the recovered signal is passed through an envelope detector to recover the original signal that was transmitted

2.3.5   AM Demodulation

Once the modulated analog signal has been transmitted, it needs to be received and demodulated. This is accomplished by the use of a band-pass filter that is tuned to the appropriate carrier frequency. Figure 2.7 shows the conceptual model of the AM receiver. As the signal enters the receiver, it passes through the band-pass filter, which is tuned to the carrier frequency f_0. Next, the recovered signal is passed through an envelope detector to recover the original signal that was transmitted.

2.3.6   Drawbacks in AM

- The modulated signal contains the carrier; carrier takes power and it does not have the information
- Therefore, AM is inefficient in power usage
- Moreover, there are two sidebands, containing the same information
- It is bandwidth inefficient
- AM is also susceptible to interference, since it affects the amplitude of the carrier.

Therefore, a solution is needed to improve bandwidth and power efficiency.

Problem 2.1 Given:

- Input modulating frequency f_m = 10 kHz
- Carrier frequency f_c = 400 kHz

Find

- Spectral components
- Bandwidth

Solution

Spectral Components:

- f_c = 400 kHz
- $f_c + f_m$ = 400 kHz + 10 kHz = 410 kHz
- $f_c - f_m$ = 400 kHz − 10 kHz = 390 kHz

Bandwidth

- BW = $2 f_m$ = 2 × 10 kHz = 20 kHz.

2.4   Double Sideband-Suppressed Carrier (DSBSC)

2.4.1   DSBSC Modulation

Double sideband-suppressed carrier (DSBSC), also known as product modulator, is an AM signal that has a suppressed carrier [3–5]. Let us take the original AM signal once again, as given below:

$$S(t) = A_c \cos(2\pi f_c t) + (1/2) A_c A_m \cos[2\pi(f_c + f_m)t] \\ + (1/2) A_c A_m \cos[2\pi(f_c - f_m)t] \tag{2.13}$$

Notice that there are three spectral components:

- The first term is the carrier only, which does not have any information
- The second and third terms contain information.

In DSBSC, we suppress the carrier, which is the first term that does not have any information. Therefore, by suppressing the first term we obtain the following:

$$S(t) = (1/2) A_c A_m \cos[2\pi(f_c + f_m)t] + (1/2) A_c A_m \cos[2\pi(f_c - f_m)t] \tag{2.14}$$

Next, we use the following trigonometric identities:

- $\cos(A + B) = \cos A \cos B - \sin A \sin B$
- $\cos(A - B) = \cos A \cos B + \sin A \sin B$

With $A = 2\pi f_m t = \omega_m t$ and $B = 2\pi f_c t = \omega_c t$, we obtain:

$$S(t) = A_c A_m \cos(\omega_m t) \cos(\omega_c t) \tag{2.15}$$

Now, define

- $m(t) = A_m \cos(\omega_m t)$
- $C(t) = A_c \cos(\omega_c t)$

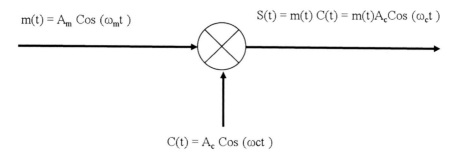

Fig. 2.8 Symbolic representation of DSBSC, also known as product modulator

Then, we can write the above equations as:

$$S(t) = m(t)\, C(t) \qquad\qquad (2.16)$$

This is the DSBSC waveform. Since the output is the product of two signals, it is also known as product modulator. The symbolic representation is given in Fig. 2.8, where $m(t)$ is the input modulating signal and $C(t)$ is the carrier frequency.

2.4.2   Generation of DSBSC Signal

A DSBSC signal can be generated using two AM modulators arranged in a balanced configuration as shown in Fig. 2.9 [3–5]. The outcome is a cancellation of the discrete carrier. Also, the output is the product of two inputs: $S(t) = m(t)\, C(t)$. This is why it is called "product modulator."

Proof of DSBSC

Consider the DSBSC modulator as shown in Fig. 2.9. Here, the AM modulators generate $S_1(t)$ and $S_2(t)$, which are given by:

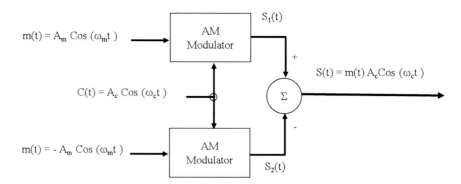

Fig. 2.9 Construction of DSBSC modulator. The output is the product of two signals

$$S_1(t) = A_c[1 + m(t)] \cos(\omega_c t) \tag{2.17}$$

$$S_2(t) = A_c[1 - m(t)] \cos(\omega_c t)$$

Subtracting $S_2(t)$ from $S_1(t)$, we essentially cancel the carrier to obtain:

$$\begin{aligned} S(t) &= S_1(t) - S_2(t) \\ &= 2\, m(t)\, A_c\, \cos(\omega_c t) \end{aligned} \tag{2.18}$$

Therefore, except for the scaling factor 2, the above equation is exactly the same as the desired DSBSC waveform shown earlier, which does not have the carrier. In other words, the carrier has been suppressed, hence the name double sideband-suppressed carrier (DSBSC).

2.4.3   DSBSC Spectrum and Bandwidth

We begin with the DSBSC-modulated signal:

$$S(t) = 2\, m(t)\, A_c \cos(\omega_c t) \tag{2.19}$$

where

$$m(t) = A_m \cos(\omega_m t)$$

Therefore,

$$S(t) = 2 A_c A_m \cos(\omega_m t)\, \cos(\omega_c t) \tag{2.20}$$

This is the desired DSBSC waveform for spectral analysis.
Now, use the trigonometric identity:

- $\cos A\, \cos B = 1/2 \cos(A + B) + 1/2 \cos(A - B) \tag{2.21}$

where

$$A = \omega_m t = 2\pi f_m t \text{ and } B = \omega_c t = 2\pi f_c t$$

Therefore,

$$\begin{aligned} S(t) &= A_c[\cos(\omega_c + \omega_m)t + \cos(\omega_c - \omega_m)t] \\ &= A_c[\cos 2\pi(f_c + f_m)t + \cos 2\pi(f_c - f_m)t] \end{aligned} \tag{2.22}$$

Fig. 2.10 DSBSC spectrum where the carrier frequency is suppressed. The bandwidth is given by $2 f_m$

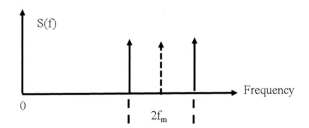

Notice that the carrier power is distributed among the sidebands (Fig. 2.10). Therefore, it is more efficient. The bandwidth is given by:

$$BW = 2f_m \tag{2.23}$$

2.4.4 DSBSC Drawbacks

- There are two identical sidebands.
- Each sideband contains the same information
- Bandwidth is $2 f_m$
- Unnecessary power usage

Therefore, a solution is needed to improve bandwidth efficiency.

Problem 2.2 Given:Two product modulators using identical carriers are connected in a series, as shown below:

$m(t) = A_m \, \text{Cos}(\omega_m t)$ $S_1(t)$ $S_2(t)$

$C(t) = A_c \, \text{Cos} \, (\omega_c t)$ $C(t) = A_c \, \text{Cos} \, (\omega_c t)$

Find

(a) The output waveform $S_2(t)$
(b) What is the function of this circuit?

Solution

(a)
$$S_1(t) = m(t) \, C(t)$$
$$S_2(t) = S_1(t) \, C(t) = m(t)C^2(t) = A_m \cos(\omega_m t) \, Ac^2 \cos 2(\omega_c t)$$

(b) The function of the circuit is to demodulate DSBSC signals, where the carrier frequency is filtered out.

2.5   Single Sideband (SSB) Modulation

2.5.1   Why SSB Modulation?

- The basic AM has a carrier which does not carry information—*Inefficient power usage*
- The basic AM has two sidebands contain the same information—*Additional loss of power*
- DSBSC has two sidebands, containing the same information—*Loss of power*
- Therefore, the basic AM and DSBSC are bandwidth and power inefficient
- SSB is bandwidth and power efficient.

2.5.2   Generation of SSB-Modulated Signal

Single sideband (SSB) modulation uses two product modulators as shown in Fig. 2.11 [3–5], where

$$\bullet \quad m(t) = A_m \cos(\omega_m t) \tag{2.24}$$

$$\bullet \quad m(t)^* = A_m \sin(\omega_m t) - (\text{Hilbert Transform}) \tag{2.25}$$

$$\bullet \quad C(t) = A_m \cos(\omega_c t) \tag{2.26}$$

$$\bullet \quad C(t)^* = A_c \sin(\omega_c t) - (\text{Hilbert Transform}) \tag{2.27}$$

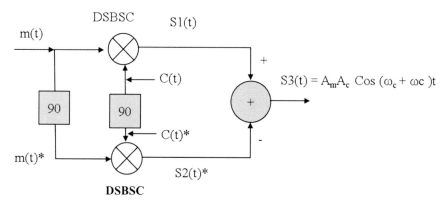

Fig. 2.11 Generation of SSB signal

Solving for S_1, S_2, and S_3, we obtain:

- $S_1(t) = A_c A_m \cos(\omega_m t) \cos(\omega_c t)$ (2.28)

- $S_2(t) = A_c A_m \sin(\omega_m t) \sin(\omega_c t)$ (2.29)

$$S_3(t) = S_1(t) - S_2(t)$$
$$= A_c A_m \cos(\omega_m t) \cos(\omega_c t) - A_c A_m \sin(\omega_m t) \sin(\omega_c t) \tag{2.30}$$

Using the following formula:

- $\cos A \cos B = 1/2\cos(A + B) + 1/2\cos(A - B)$
- $\sin A \sin B = 1/2\cos(A - B) - 1/2\cos(A + B)$

Solving for S_3, we get:

$$S_3 = A_c A_m \cos(\omega_c + \omega_m)t \tag{2.31}$$

In the above equation, $S_3(t)$ is the desired SSB signal, which is the upper sideband only.

2.5.3 SSB Spectrum and Bandwidth

Let us consider the SSB signal again:

$$S_3 = A_c A_m \cos(\omega_c + \omega_m)t$$
$$= A_c A_m \cos 2\pi(f_c + f_m)t \tag{2.32}$$

Here, we see that the SSB spectrum contains only one sideband. Therefore, it is more efficient. The SSB bandwidth is given by:

$$\text{SSB BW} = f_m \tag{2.33}$$

Figure 2.12 displays the SSB spectrum.

Fig. 2.12 SSB spectrum showing the upper sideband. The SSB bandwidth is f_m

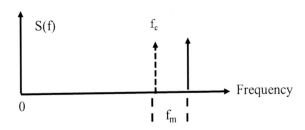

Problem 2.3
Given:

- $m(t) = A_m \cos(\omega_m t)$
- $m(t)^* = A_m \sin(\omega_m t)$ − (Hilbert Transform)
- $C(t) = A_c \cos(\omega_c t)$
- $C(t)^* = A_c \sin(\omega_c t)$ − (Hilbert Transform)

Design an SSB modulator to realize the lower sideband. Sketch the spectral response.

Solution
Solving for S_1 and S_2, we obtain:

- $S_1(t) = A_c A_m \cos(\omega_m t) \cos(\omega_c t)$
- $S_2(t) = A_c A_m \sin(\omega_m t) \sin(\omega_c t)$

Obtain S_3 as:

$$S_3(t) = S_1(t) + S_2(t)$$
$$= A_c A_m \cos(\omega_m t)\cos(\omega_c t) + A_c A_m \sin(\omega_m t)\sin(\omega_c t)$$

Using the following formula:

- $\cos A \cos B = 1/2\cos(A + B) + 1/2\cos(A − B)$
- $\sin A \sin B = 1/2\cos(A − B) − 1/2\cos(A + B)$

Solving for S_3, we get:

$$S_3 = A_c A_m \cos(\omega_m − \omega_c)t$$
$$= A_c A_m \cos 2\pi(f_c − f_m)t$$

In the above equation, $S_3(t)$ is the desired SSB signal, which is the lower sideband only. The spectral response, showing the lower sideband, is presented below.

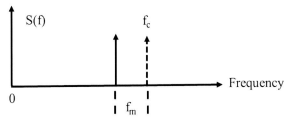

2.6   Conclusions

This chapter presents the key concepts and underlying principles of Amplitude Modulation. It was shown how the audio waveform changes the amplitude of the carrier to determine the envelope of the modulated carrier. It was also shown that the modulated carrier contains several spectral components that lead to DSBSC and SSB modulation techniques. In particular, the following topics were presented in this chapter:

- Amplitude Modulation (AM)
- AM spectrum and bandwidth
- Double sideband-suppressed carrier (DSBSC)
- DSBSC spectrum and bandwidth
- Single sideband (SSB)
- SSB spectrum and bandwidth

These materials have been augmented by diagrams and associated waveforms to make them easier for readers to grasp.

References

1. Marconi G (1897) Improvements in transmitting electrical impulses and signals, and in apparatus therefor. British patent No. 12,039. Date of Application 2 June 1896; Complete Specification Left, 2 Mar 1897; Accepted, 2 July 1897 (later claimed by Oliver Lodge to contain his own ideas which he failed to patent)
2. Marconi G (1900) Improvements in apparatus for wireless telegraphy. British patent No. 7,777. Date of Application 26 Apr 1900; Complete Specification Left, 25 Feb 1901; Accepted, 13 Apr 1901
3. Leon W, Couch II (2001) Digital and analog communication systems, 7th edn. Prentice-Hall Inc, Englewood Cliffs. ISBN 0-13-142492-0
4. Godse AP, Bakshi UA (2009) Communication engineering. Technical Publications, p 36. ISBN 978-81-8431-089-4
5. Silver W (ed) (2011) Chapter 14 transceivers. The ARRL handbook for radio communications, 88th edn. American Radio Relay League. ISBN 978-0-87259-096-0

Chapter 3
Frequency Modulation (FM)

Topics

- Introduction
- Frequency Modulation (FM)
- FM Spectrum
- Carson's Rule & FM Bandwidth
- Bessel Function & FM Bandwidth
- FM bandwidth Dilemma

3.1 Introduction

In Frequency Modulation (FM), the frequency of the carrier changes in accordance with the input analog signal, while the amplitude of the carrier remains the same [1–5]. This is shown in Fig. 3.1, where

- $m(t)$ is the input modulating audio signal,
- $C(t)$ is the carrier frequency, and
- $S(t)$ is the FM-modulated carrier frequency.

As shown in the figure, the audio waveform changes the frequency of the carrier. This enables the receiver to extract the audio signal by demodulation. Notice that the frequency of the carrier changes in accordance with the input signal, while the amplitude of the carrier does not change after modulation. However, it can be shown that the modulated carrier $S(t)$ contains an infinite number of spectral components, requiring frequency domain analysis [3]. In an effort to examine this, this chapter will present the following topics:

- The basic Frequency Modulation (FM),
- FM spectrum, and
- FM bandwidth.

© The Author(s) 2017   33
S. Faruque, *Radio Frequency Modulation Made Easy*,
SpringerBriefs in Electrical and Computer Engineering,
DOI 10.1007/978-3-319-41202-3_3

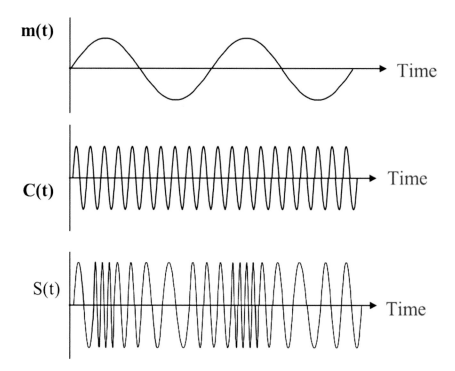

Fig. 3.1 Illustration of FM

3.2 Frequency Modulation (FM)

3.2.1 Background

FM is a form of angle modulation, where the frequency of the carrier varies in accordance with the input signal. Here, the angle refers to the angular frequency (ω). The angular frequency ω is also recognized as angular speed or circular frequency. It is a measure of rotation rate or the rate of change of the phase of a sinusoidal waveform as illustrated in Fig. 3.2.

Magnitude of the angular frequency ω is defined by one revolution or 2π radians:

$$\omega = 2\pi f = d\theta/dt \quad \text{Radians per second} \tag{3.1}$$

where,

- ω = Angular frequency in radians per seconds,
- f = Frequency in Hertz (Hz) or cycles per second, and
- θ = Phase angle.

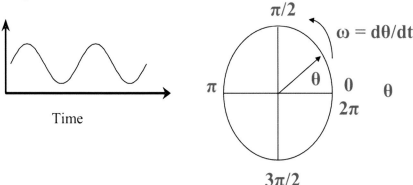

Fig. 3.2 A sinusoidal waveform in the time domain and its representation in the phase domain

Notice that in Eq. 3.1, the angular frequency ω is greater than the frequency f by a factor of 2π. Now solving for the phase angle, we obtain,

$$\theta_i(t) = 2\pi \int_0^t f_i dt \tag{3.2}$$

- θ_i = Instantaneous phase angle and
- f_i = Instantaneous frequency.

This forms the basis of our derivation of FM as presented in the following section.

3.2.2 The Basic FM

Frequency Modulation (FM) is a method of analog modulation that utilizes frequency variation of the relative frequency of the career [1]. The signal to be modulated and transmitted is analog. This is referred to as FM, where the frequency of the carrier changes in accordance with the input signal. The modulated carrier frequency f_c varies back and forth and depends on amplitude A_m and frequency f_m of the input signal.

Figure 3.3 shows the functional diagram of a typical FM, using a single-tone modulating signal. Here, $m(t)$ is the input analog signal we want to transmit, $C(t)$ is the carrier frequency without modulation, and $S(t)$ is the output FM-modulated carrier frequency. These parameters are described below.

m(t) = A_m Cos ($\omega_m t$)

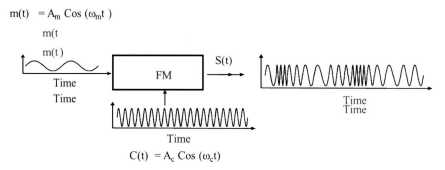

Fig. 3.3 A functional diagram of a typical FM modulator using a single-tone input modulating signal. Here, $m(t)$ is the input analog signal we want to transmit, $C(t)$ is the carrier frequency without modulation, and $S(t)$ is the output FM-modulated carrier frequency

We examine this by means of a single-tone input modulating signal and its angular frequency as shown in Fig. 3.2. Since the frequency of the carrier varies in accordance with the input signal, the instantaneous frequency of the carrier is given by

$$f_i(t) = f_c(t) + D_f m(t) \tag{3.3}$$

where,

- f_i = Instantaneous frequency,
- f_c = Carrier frequency,
- D_f = Constant, and
- $m(t) = A_m \cos(wmt)$.

The FM-modulated signal is given by

$$S(t) = A_c \cos(\theta_i) \tag{3.4}$$

where Ac is the amplitude of the carrier frequency and θ_i is the instantaneous angle. Substituting Eq. (3.2) into Eq. (3.4) for θi, we get

$$S(t) = A_c \cos[2\pi \int_0^t f_i dt] \tag{3.5}$$

where f_i is the instantaneous frequency. Substituting Eq. (3.3) into Eq. (3.5) for fi, we obtain

$$S(t) = A_c \cos[2\pi \int_0^t \{fc(t) + Df\, m(t)\}dt] \tag{3.6}$$

where $m(t) = A_m \cos(\omega_m t)$ is the input modulating signal. Integrating the above equation, we obtain the desired FM signal as follows:

$$S(t) = A_c \cos\left[2\pi f_c t + \left(\frac{\Delta f}{f_m}\right)\sin(2\pi f_m t)\right]$$

$$= A_c \cos[2\pi f_c t + \beta \sin(2\pi f_m t)] \qquad (3.7)$$

$$\beta = \left(\frac{\Delta f}{f_m}\right) = \text{Modulation Index} \quad \Delta f = DfAm = \text{Freq. Deviation}$$

where,

- $S(t)$ = FM-modulated carrier signal,
- f_c = Frequency of the carrier,
- A_c = Amplitude of the carrier frequency,
- $\Delta f = D_f A_m$ = Frequency deviation,
- f_m = Input modulating frequency,
- Am = Amplitude of the input modulating signal,
- D_f = A constant parameter, and
- $\beta = \Delta f / f_m$ = Modulation index.

Note that the modulation index β is an important design parameter in FM. It is directly related to FM bandwidth. It may also be noted that FM bandwidth depends on both frequency and amplitude of the input modulating signal. Let us take a closer look.

3.3   FM Spectrum and Bandwidth

In wireless communications, the scarcity of RF spectrum is well known. For this reason, we have to be vigilant about using transmission bandwidth and modulation. The transmission bandwidth depends on the following:

- Spectral response of the input modulating signal,
- Spectral response of the carrier frequency, and
- Modulation type.

Let's take a closer look!

3.3.1   Spectral Response of the Input Modulating Signal

In FM, the input modulating signal is a continuous time low-frequency analog signal. For simplicity, we use a sinusoidal waveform, which is periodic and

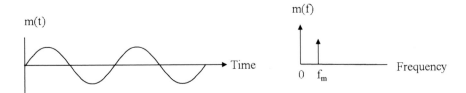

Fig. 3.4 A low-frequency sine wave and its frequency response

continuous with respect to time. It has one frequency component. For example, the sine wave is described by the following time domain equation:

$$V(t) = V_p \sin(\omega_m t) \qquad (3.8)$$

where,

- Vp = Peak voltage,
- $\omega_m = 2\pi f_m$, and
- f_m = Input modulating frequency in Hz.

Figure 3.4 shows the characteristics of a sine wave and its spectral response. Since the frequency is constant, its spectral response is located in the horizontal axis at f_m and the peak voltage is shown in the vertical axis. The corresponding bandwidth is zero.

3.3.2 Spectral Response of the Carrier Frequency

A carrier frequency (f_c) is essentially a sinusoidal waveform, which is periodic and continuous with respect to time. It has one frequency component, which is much higher than the input modulating frequency $(f_c \gg f_m)$. For example, the sine wave is described by the following time domain equation:

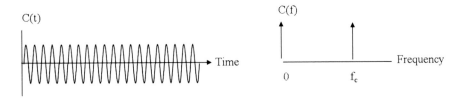

Fig. 3.5 A high-frequency sine wave and its frequency response before modulation

$$V(t) = V_p \sin(\omega_c t) \tag{3.9}$$

where
$$V_p = \text{Peak voltage}$$

- $\omega_c = 2\pi f_c$ and
- f_c = Carrier frequency in Hz.

Figure 3.5 shows the characteristics of a high-frequency sine wave and its spectral response. Since the frequency is constant, its spectral response is located in the horizontal axis at f_c and the peak voltage is shown in the vertical axis. The corresponding bandwidth is zero.

3.3.3 FM Spectrum

In FM, the frequency of the carrier changes in accordance with the input signal. Here, we have:

- Input Signal : $m(t) = A_m \cos(\omega_m t)$
- Carrier Frequency : $C(t) = A_c \cos(\omega_c t)$ (3.10)
- Modulated Carrier : $S(t) = A_c \cos[(\omega_c t) + \beta \sin(\omega_m t)]$

where

- $S(t)$ = The modulated carrier,
- Ac = Frequency of the carrier,
- ω_c = Nominal frequency of the carrier frequency,
- $\beta = \Delta f / fm$ = Modulation index,
- $\Delta f = D_f A_m$ = Frequency deviation,
- D_f = Constant, and
- A_m = Amplitude of the input modulating signal.

By inspecting the modulated carrier frequency, we observe that $S(t)$ depends on both frequency and amplitude of the input signal. The spectral response is given in Fig. 3.6. Notice that the carrier frequency after modulation varies back and forth from the nominal frequency f_c as depicted in the figure. In Fig. 3.6, we see that:

- As time passes, the carrier moves back and forth in frequency in exact step with the input signal.
- Frequency deviation is proportional to the input signal voltage.
- A group of many sidebands is created, spaced from carrier by amounts $N \times f_i$.
- Relative strength of each sideband depends on Bessel function.
- Strength of individual sidebands far away from the carrier is proportional to (freq. deviation $\times$ input frequency).
- Higher order spectral components are negligible.

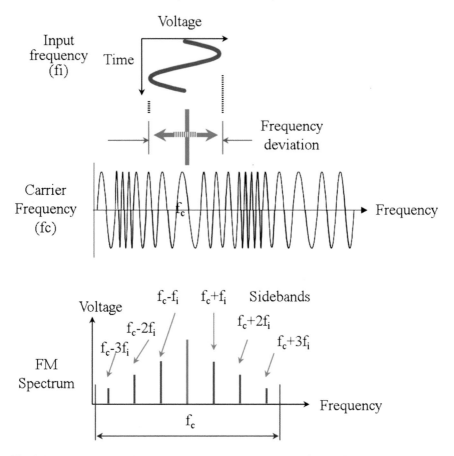

Fig. 3.6 FM spectrum. As time passes, the carrier moves back and forth in frequency in exact step with the input signal and generates an infinite number of sidebands

- Carson's rule can be used to determine the approximate bandwidth: bandwidth required = 2 × (highest input frequency + frequency deviation).

3.3.4 Carson's Rule and FM Bandwidth

The Carson's rule, a rule of thumb, states that more than 98 % of the power of FM signal lies within a bandwidth given by the following approximation:

$$\text{FM Bandwidth (BW)} = 2f_m(1 + \beta) \tag{3.11}$$

where

Table 3.1 Bessel function table

β	J0	J1	J2	J3	J4	J5
0	1					
0.25	0.98	0.12				
0.5	0.94	0.24	0.03			
1	0.77	0.44	0.11	0.02		
1.5	0.51	0.56	0.23	0.06	0.01	
2	0.22	0.58	0.35	0.13	0.03	

- $\beta = \Delta f / fm$ = Modulation index,
- Δf = Peak deviation of the instantaneous frequency from the center of the carrier frequency, and
- fm = Highest frequency of the modulating signal.

3.3.5 Bessel Function and FM Bandwidth

FM bandwidth can be estimated by means of Bessel function of the first kind. For a single-tone modulation, it can be obtained as a function of the sideband number and the modulation index.

For a given β, the solution for $S(t)$ is given as follows:

$$
\begin{aligned}
S(t) = {} & A_c \cos[(\omega_c t) + \beta \sin(\omega_m t)] \\
= {} & J_0(\beta)\cos(\omega_c t) \\
& + J_1(\beta)\cos(\omega_c t + \omega_m t) + J_2(\beta)\cos(\omega_c t + 2\omega_m t) + J_3(\beta)\cos(\omega_c t + 3\omega_m t) + \ldots \\
& - J_1(\beta)\cos(\omega_c t - \omega_m t) - J_2(\beta)\cos(\omega_c t - 2\omega_m t) - J_3(\beta)\cos(\omega_c t - 3\omega_m t) + \ldots
\end{aligned}
$$

$$(3.12)$$

Here, J's are the Bessel functions, representing the amplitude of the sidebands. $J_0(\beta)$ is the amplitude of the fundamental spectral component, and the remaining spectral components are the sidebands. Each sideband is separated by the input modulating frequency. These values are also available as a standard Bessel function table.

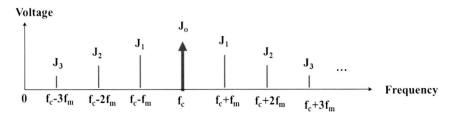

Fig. 3.7 FM sidebands. Power is taken from the carrier $J0$ and distributed among the sidebands $J1$, $J2$, $J3$,..., etc. Each sideband is separated by the modulating frequency f_m

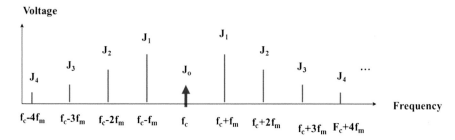

Fig. 3.8 FM sidebands for large β. More power is taken from the carrier $J0$ and distributed among the sidebands $J1$, $J2$, $J3$, $J4$,... Each sideband is separated by the modulating frequency f_m

As an example, Table 3.1 provides a few Bessel parameters to illustrate the concept. In this table, the carrier and sideband amplitude powers, $J0$, $J1$, $J2$,..., etc., are presented for different values of β. Here, $J0$ is carrier power before modulation. After modulation, with a given value of β, power is taken from the carrier and distributed among the sidebands. Also, there is a unique value of β, for which the carrier amplitude becomes zero and all the signal power is in the sidebands.

Note that, in FM, the sidebands are on both sides of the carrier. Therefore, the total bandwidth includes spectral components from both sides of the carrier. For a given low β, this is shown in Fig. 3.7. Here, J0 is the spectral component of the carrier. $J1$, $J2$, $J3$, etc., are the sidebands. Each sideband is separated by the input modulating frequency f_m. After modulation, power is taken from the carrier J0 and distributed among the sidebands, depending on the modulation index β.

Figure 3.8 shows another scenario where β is large. In this case, more power is taken from the carrier and distributed among the sideband, while creating more significant sidebands, requiring more bandwidth.

3.3.6 FM Bandwidth Dilemma

In FM, we notice that if the modulation index is low, the occupied bandwidth is low and the sidebands take less power from the carrier, making the modulation less efficient. On the other hand, if the modulation index is high, the occupied bandwidth is also high, while the sidebands retain most of the power, making the modulation more efficient at the expense of bandwidth. This is a dilemma in FM. However, FM is more popular because it is less sensitive to noise.

Problem 3.1
Given:

$$f_m = 1\,\text{kHz}$$
$$f_c = 1\,\text{MHz}$$
$$\beta = 1$$

Find:

(a)   The FM bandwidth using Carson's rule and
(b)   The FM bandwidth using Bessel function

Solution:

(a)   FM BW (Carson's rule) = $2 f_m (1 + \beta) = 2 \times 1$ kHz $(1 + 1) = 4$ kHz.
(b)   For $\beta = 1$, See Table 3.1:

- $J0 = 0.77$, $J1 = 0.44$, $J2 = 0.11$, $J3 = 0.02$

where $J0$ is the carrier and $J1$, $J2$, and $J3$ are the sidebands. Here, we can neglect $J3$ since it has a negligible power. Therefore, there are two significant upper and lower sidebands, and each sideband is 1 kHz apart. Therefore, FM bandwidth is $2 \times$ Number of significant sidebands $= 2 \times 2 = 4$ kHz. See figure below.

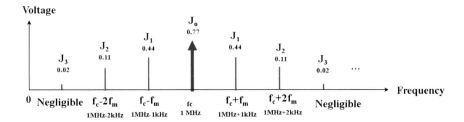

Problem 3.2
Given:

$$fm = 1 \text{ kHz}$$
$$fc = 1 \text{ MHz}$$
$$\beta = 2$$

Find:

(a)   The FM bandwidth using Carson's rule and
(b)   The FM bandwidth using Bessel function

Solution:

(a)   FM BW (Carson's rule) = $2 f_m (1 + \beta) = 2 \times 1$ kHz $(1 + 2) = 6$ kHz.
(b)   For $\beta = 2$, See Table 3.1:

- $J0 = 0.22$, $J1 = 0.58$, $J2 = 0.35$, $J3 = 0.13$, $J4 = 0.03$

Here, we have 4 upper and 4 lower sidebands, and each sideband is separated by 1 kHz. Since the J4 is negligible, we take 3 upper sidebands and 3 lower sidebands. Therefore, the FM bandwidth is $2 \times 3 = 6$ kHz. See figure below.

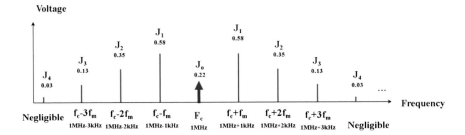

3.4   Conclusions

- This chapter presents a brief overview of Frequency Modulation (FM) and its attributes.
- The key concept and the underlying principle of FM are presented with numerous illustrations to make them easier for readers to grasp.
- FM spectrum is addressed with illustrations.
- FM bandwidth is estimated by means of Carson's rule and Bessel function.
- FM bandwidth dilemma is explained, and problems are given to illustrate the concept.

References

1. Armstrong EH (1936) A method of reducing disturbances in radio signaling by a system of frequency modulation. Proc IRE (IRE) 24(5):689–740. doi:10.1109/JRPROC.1936.227383
2. Lathi BP (1968) Communication systems. Wiley, New York, pp 214–217. ISBN 0-471-51832-8
3. Leon W, Couch II (2001) Digital and analog communication systems, 7th edn. Prentice-Hall Inc, Englewood Cliffs. ISBN 0-13-142492-0
4. Haykin S (2001) Communication systems, 4th edn. Wiley, New York
5. Felix MO (1965) FM systems of exceptional bandwidth. Proc IEEE 112(9):1664

Chapter 4
Amplitude Shift Keying (ASK)

Topics

- Introduction
- Amplitude Shift Keying (ASK)
- ASK spectrum
- ASK Bandwidth
- Performance Analysis

4.1 Introduction

In amplitude shift keying (ASK), the amplitude of the carrier changes in discrete levels in accordance with the input digital signal, while the frequency of the carrier remains the same. This is shown in Fig. 4.1, where

- $m(t)$ is the input modulating digital signal,
- $C(t)$ is the carrier frequency, and
- $S(t)$ is the ASK-modulated carrier frequency.

As shown in the figure, the digital binary signal changes the amplitude of the carrier in two discrete levels. This enables the receiver to extract the digital signal by demodulation. Notice that the amplitude of the carrier changes in accordance with the input signal, while the frequency of the carrier does not change after modulation. However, it can be shown that the modulated carrier $S(t)$ contains several spectral components, requiring frequency domain analysis.

In the following sections, the above disciplines in ASK modulation will be presented along with the respective spectrum and bandwidth. These materials have been augmented by diagrams and associated waveforms to make them easier for readers to grasp.

© The Author(s) 2017
S. Faruque, *Radio Frequency Modulation Made Easy*,
SpringerBriefs in Electrical and Computer Engineering,
DOI 10.1007/978-3-319-41202-3_4

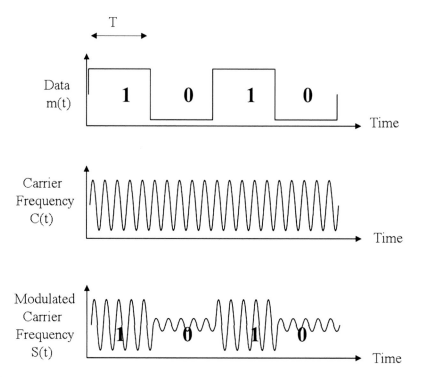

Fig. 4.1 ASK waveforms. The amplitude of the carrier changes in accordance with the input digital signal. The frequency of the carrier remains the same

4.2 ASK Modulation

Amplitude shift keying (ASK), also known as on–off keying (OOK), is a method of digital modulation that utilizes amplitude shifting of the relative amplitude of the career frequency [1–3]. The signal to be modulated and transmitted is binary, which is encoded before modulation. This is an indispensable task in digital communications, where redundant bits are added with the raw data that enables the receiver to detect and correct bit errors, if they occur during transmission [4–16].

While there are many error-coding schemes available, we will use a simple coding technique, known as "Block Coding" to illustrate the concept.

Figure 4.2 shows an encoded ASK modulation scheme using (15, 8) block code where an 8-bit data block is formed as M-rows and N-columns ($M = 4$, $N = 2$). The product $MN = k = 8$ is the dimension of the information bits before coding. Next, a horizontal parity P_H is appended to each row and a vertical parity P_V is appended to each column. The resulting augmented dimension is given by the product $(M + 1)(N + 1) = n = 15$, which is then ASK modulated and transmitted row by row. The rate of this coding scheme is given by

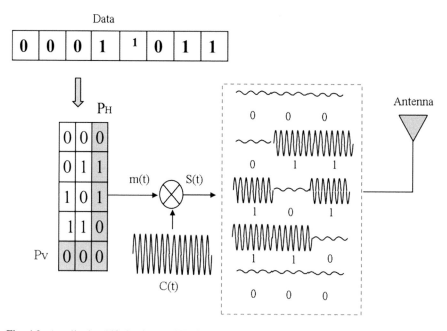

Fig. 4.2 Amplitude shift keying (ASK) is also known as on–off keying (OOK). The input encoded data block is transmitted row by row. The amplitude of the carrier frequency changes in accordance with the input digital signal

$$\text{Code Rate: } r = (MN)/[(M+1)(N+1)] = (4 \times 2)/(5 \times 3) = 8/15 \qquad (4.1)$$

The coded bit rate R_{b2} is given by

$$R_{b2} = \text{Uncoded Bit Rate/Code Rate} = R_{b1}/r = R_{b1} \,(15/8) \qquad (4.2)$$

Next, the coded bits are modulated by means of the ASK modulator as shown in the figure. Here,

- The Input digital signal is the encoded bit sequence we want to transmit
- Carrier is the radio frequency without modulation
- Output is the ASK-modulated carrier, which has two amplitudes corresponding to the binary input signal. For binary signal 1, the carrier is ON. For the binary signal 0, the carrier is OFF; however, a small residual signal may remain due to noise, interference, etc., as indicated in the figure.

As shown in the figure, the amplitude of the carrier changes in discrete levels, in accordance with the input signal, where

- Input Data : $m(t) = 0$ or 1 (coded data)
- Carrier Frequency : $C(t) = A \cos(\omega t)$ (4.3)
- Modulated Carrier : $S(t) = m(t)C(t) = m(t)A \cos(\omega t)$

Therefore,

$$\text{For } m(t) = 1 : S(t) = A \cos(\omega t) \text{ i.e. the carrier is ON}$$
$$\text{For } m(t) = 1 : S(t) = 0, \text{ i.e. the carrier is OFF}$$ (4.4)

where A is the amplitude and ω is the frequency of the carrier.

4.3   ASK Demodulation

Once the modulated binary data has been transmitted, it needs to be received and demodulated. This is often accomplished with the use of a band-pass filter. In the case of ASK, the receiver needs to utilize one band-pass filter that is tuned to the appropriate carrier frequency. As the signal enters the receiver, it passes through the filter and a decision as to the value of each bit is made to recover the encoded data block, along with horizontal and vertical parities. Next, the receiver appends horizontal and vertical parities $P_H{}^*$ and $P_V{}^*$ to check parity failures and recovers the data block. This is shown in Fig. 4.3 having no errors. If there is an error, there will be a parity failure in $P_H{}^*$ and $P_V{}^*$ to pinpoint the error.

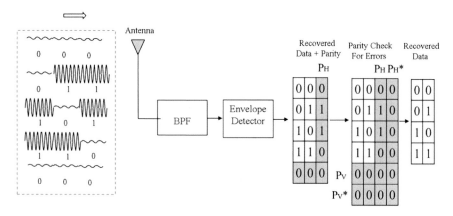

Fig. 4.3 Data recovery process in ASK, showing no errors. If there is an error, there will be a parity failure in $P_H{}^*$ and $P_V{}^*$ to pinpoint the error

4.4 ASK Bandwidth

In wireless communications, the scarcity of RF spectrum is well known. For this reason, we have to be vigilant about using transmission bandwidth in error control coding and modulation. The transmission bandwidth depends on:

- Spectral response of the encoded data
- Spectral response of the carrier frequency, and
- Modulation type.

4.4.1 Spectral Response of the Encoded Data

In digital communications, data is generally referred to as a non-periodic digital signal. It has two values:

- Binary-1 = High, Period = T
- Binary-0 = Low, Period = T

Also, data can be represented in two ways:

- Time domain representation and
- Frequency domain representation

The time domain representation (Fig. 4.4a), known as non-return-to-zero (NRZ), is given by:

$$V(t) = V \qquad < 0 < t < T$$
$$= 0 \qquad \text{elsewhere} \tag{4.5}$$

The frequency domain representation is given by "Fourier transform":

$$V(\omega) = \int_0^T V \cdot e^{-j\omega t} dt \tag{4.6}$$

$$|V(\omega)| = VT \left[\frac{\sin(\omega T/2)}{\omega T/2} \right]$$

$$P(\omega) = \left(\frac{1}{T} \right) |V(\omega)|^2 = V^2 T \left[\frac{\sin(\omega T/2)}{\omega T/2} \right]^2 \tag{4.7}$$

Here, $P(\omega)$ is the power spectral density. This is plotted in Fig. 4.4b. The main lobe corresponds to the fundamental frequency and side lobes correspond to harmonic components. The bandwidth of the power spectrum is proportional to the frequency. In practice, the side lobes are filtered out since they are relatively insignificant with respect to the main lobe. Therefore, the one-sided bandwidth is given by the ratio $f/f_b = 1$. In other words, the one-sided bandwidth $= f = f_b$, where $f_b = R_b = 1/T$, T being the bit duration.

The general equation for two-sided response is given by:

$$V(\omega) = \int_{-\infty}^{\infty} V(t) \cdot e^{-j\omega t} dt$$

In this case, $V(\omega)$ is called two-sided spectrum of $V(t)$. This is due to both positive and negative frequencies used in the integral. The function can be a voltage or a current. Figure 4.4c shows the two-sided response, where the bandwidth is determined by the main lobe as shown below:

Two-sided bandwidth (BW) $= 2R_b$ ($R_b =$ Bit rate before coding)   (4.8)

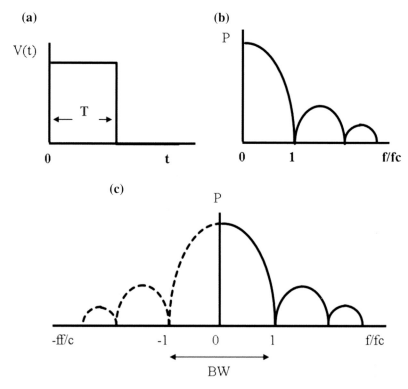

Fig. 4.4 **a** Discrete time digital signal **b** it is one-sided power spectral density and **c** two-sided power spectral density. The bandwidth associated with the non-return-to-zero (NRZ) data is $2R_b$, where R_b is the bit rate

Important Notes

1. If R_b is the bit rate before coding, and if the data is NRZ, then the bandwidth associated with the raw data will be $2R_b$. For example, if the bit rate before coding is 10 kb/s, then the bandwidth associated with the raw data will be 2×10 kb/s = 20 kHz.
2. If R_b is the bit rate before coding, code rate is r, and if the data is NRZ, then the bit rate after coding will be R_b (coded) = R_b (uncoded)r. The corresponding bandwidth associated with the coded data will be $2R_b$ (coded) = $2R_b$ (uncoded)/r. For example, if the bit rate before coding is 10 kb/s and the code rate $r = 1/2$, then the coded bit rate will be R_b (coded) = R_b (uncoded)/r = 10/0.5 = 20 kb/s. The corresponding bandwidth associated with the coded data will be $2 \times 20 = 40$ kHz.

4.4.2 Spectral Response of the Carrier Frequency Before Modulation

A carrier frequency is essentially a sinusoidal waveform, which is periodic and continuous with respect to time. It has one frequency component. For example, the sine wave is described by the following time domain equation:

$$V(t) = V_p \sin(\omega t_c) \qquad (4.9)$$

where

$$V_P = \text{Peak voltage}$$

- $\omega_c = 2\pi f_c$
- f_c = Carrier frequency in Hz

Figure 4.5 shows the characteristics of a sine wave and its spectral response. Since the frequency is constant, its spectral response is located in the horizontal axis and the peak voltage is shown in the vertical axis. The corresponding bandwidth is zero.

4.4.3 ASK Bandwidth at a Glance

In ASK, the amplitude of the carriers changes in discrete levels, in accordance with the input signal, where

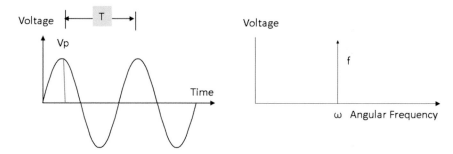

Fig. 4.5 High-frequency carrier frequency response

- Input Data : $m(t) = 0\,\text{or}\,1$
- Carrier Frequency : $C(t) = Ac\cos(\omega_c t)$
- Modulated Carrier : $S(t) = m(t)C(t) = m(t)A_c\cos(\omega_c t)$

Since $m(t)$ is the input digital signal and it contains an infinite number of harmonically related sinusoidal waveforms and that we keep the fundamental and filter out the higher-order components, we write:

$$m(t) = A_m\sin(\omega_m t)$$

The ASK-modulated signal then becomes:

$$\begin{aligned} S(t) = m(t)S(t) &= A_m A_c \sin(\omega_m t)\cos(\omega_m t) \\ &= 1/2\,A_m A_c\left[\sin(\omega_c - \omega_m)t + \sin(\omega_c + \omega_m)t\right] \end{aligned} \tag{4.10}$$

The spectral response is depicted in Fig. 3.14. Notice that the spectral response after ASK modulation is the shifted version of the NRZ data. Bandwidth is given by: BW $= 2R_b$ (coded), where R_b is the coded bit rate (Fig. 4.6).

Problem 1
Given:

- Bit rate before coding: $R_{B1} = 10$ kb/s
- Code rate: $r = 8/15$
- Modulation: ASK

Find:

(a) the bit rate after coding: R_{B2}
(b) Transmission bandwidth: BW

Solution

(a) Bit rate after coding (R_{B2})
$R_{B2} = R_{B1}/r = 10$ kb/s $(15/8) = 18.75$ kb/s
(b) Transmission BW $= 2 \times 18.75$ kb/s $= 37.5$ kHz.

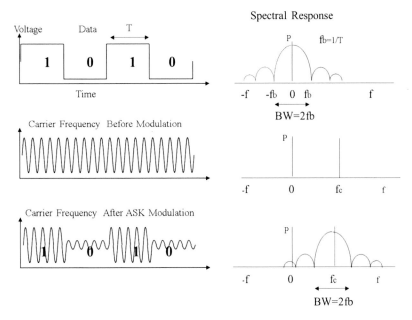

Fig. 4.6 ASK bandwidth at a glance. **a** Spectral response of NRZ data before modulation. **b** Spectral response of the carrier before modulation. **c** Spectral response of the carrier after modulation. The transmission bandwidth is $2f_b$, where f_b is the bit rate and $T = 1/f_b$ is the bit duration for NRZ data

4.5 BER Performance

It is well known that an (n, k) bock code, where k = number of information bits, n = number of coded bits, can correct t errors [4, 5]. A measure of coding gain is then obtained by comparing the uncoded word error WER_U with the coded word WER_C. We examine this by means of the following analytical means.

Let the uncoded word error be defined as (WER_U). Then, with ASK modulation, the uncoded BER will be given by:

$$BER_U = 0.5\,EXP(-E_B/2N_0) \tag{4.11}$$

The probability that the uncoded word (WER_U) will be received in error is 1 minus the product of the probabilities that each bit will be received correctly. Thus, we write:

$$WER_U = 1 - (1 - BER_U)^k \tag{4.12}$$

Let the coded word error be defined as (WER_C). Since $n > k$, the coded bit energy to noise ratio will be modified to E_c/N_0, where $E_c/N_0 = E_b/N_0 + 10\log(k/n)$. Therefore, the coded BERc will be:

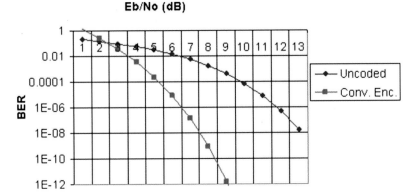

Fig. 4.7 Typical error performance in AWGN

$$\text{BER}_C = 0.5 \, \text{EXP}(-E_C/2N_0) \tag{4.13}$$

The corresponding coded word error rate is given as:

$$\text{WER}_C = \sum_{k=t+1}^{n} \binom{n}{k} \text{BER}_C^K (1 - \text{BER}_C)^{n-k} \tag{4.14}$$

When $\text{BER}_c < 0.5$, the first term in the summation is the dominant one; therefore, Eq. (4.14) can be simplified as

$$\text{WER}_C \approx \binom{n}{k} \text{BER}_c^k (1 - \text{BER}_c)^{n-k} \tag{4.15}$$

Using (15, 8) block code ($n = 15$, $k = 8$, $t = 1$), we obtain the coded and the uncoded WER as shown in Fig. 4.7. Coding gain is the difference in E_b/N_0 between the two curves. Notice that at least 3–4 db coding gain is available in this example where $r = k/n = 8/15$.

4.6 Conclusions

- This chapter presents ASK modulation and its attributes.
- Numerous illustrations are provided to show how amplitude of the carrier changes in discrete levels in accordance with the input digital signal, while the frequency of the carrier remains the same.
- The Fourier transform is used to derive the spectral components and ASK bandwidth is calculated.

- Bit error rate (BER) performance is presented.
- These materials have been augmented by diagrams and associated waveforms to make them easier for readers to grasp.

References

1. Smith DR (1985) Digital transmission systems, Van Nostrand Reinhold Co. ISBN: 0442009178
2. Leon W, Couch II (2001) Digital and analog communication systems, 7th edn. Prentice-Hall, Inc., Englewood Cliffs. ISBN: 0-13-142492-0
3. Sklar B (1988) Digital communications fundamentals and applications. Prentice Hall, Englewood Cliffs
4. Faruque S (2016) Radio frequency channel coding made easy. Springer. ISBN: 978-3-319-21169-5
5. Clark GC et al (1981) Error correction coding for digital communications. Plenum press, New York
6. Ungerboeck G (1982) Channel coding with multilevel/multiphase signals. IEEE Trans Inf Theory IT28:55–67
7. Lin S, Costello DJ Jr (1983) Error control coding: fundamentals and applications. Prentice-Hall, Inc., Englewood Cliffs
8. Blahut RE (1983) Theory and practice of error control codes. Addison-Wesley, Reading
9. van Lint JH (1992) Introduction to coding theory. GTM 86, 2nd edn. Springer, p 31. ISBN 3-540-54894-7
10. Mac Williams FJ, Sloane NJA (1977) The theory of error-correcting codes. North-Holland, p 35. ISBN 0-444-85193-3
11. Huffman W, Pless V (2003) Fundamentals of error-correcting codes. Cambridge University Press. ISBN 978-0-521-78280-7
12. Ebert PM, Tong SY (1968) Convolutional Reed-Soloman codes. Bell Syst Tech 48:729–742
13. Gallager RG (1968) Information theory and reliable communications. Wiley, New York
14. Kohlenbero A, Forney GD Jr (1968) Convolutional coding for channels with memory. IEEE Trans Inf Theory IT-14:618–626
15. Reddy SM (1968) Further results on convolutional codes derived from block codes. Inf Control 13:357–362
16. Reddy SM (1968) A class of linear convolutional codes for compound channels. Technical report, Bell Telephone Laboratories, Holmdel, New Jersey

Chapter 5
Frequency Shift Keying (FSK)

Topics

- Introduction
- Frequency Shift Keying (FSK)
- FSK spectrum
- FSK Bandwidth
- Performance Analysis

5.1 Introduction

In frequency shift keying (FSK), the frequency of the carrier changes in discrete levels in accordance with the input digital signal, while the amplitude of the carrier remains the same. This is shown in Fig. 5.1 where

- $m(t)$ is the input modulating digital signal,
- $C(t)$ is the carrier frequency, and
- $S(t)$ is the FSK-modulated carrier frequency.

As shown in the figure, the digital binary signal changes the frequency of the carrier on two discrete levels. This enables the receiver to extract the digital signal by demodulation. Notice that the frequency of the carrier changes in accordance with the input signal, while the amplitude of the carrier does not change after modulation. However, it can be shown that the modulated carrier $S(t)$ contains several spectral components, requiring frequency-domain analysis.

In the following sections, the above disciplines in FSK modulation will be presented, along with the respective spectrum and bandwidth. These materials have been augmented by diagrams and associated waveforms to make them easier for readers to grasp.

© The Author(s) 2017   57
S. Faruque, *Radio Frequency Modulation Made Easy*,
SpringerBriefs in Electrical and Computer Engineering,
DOI 10.1007/978-3-319-41202-3_5

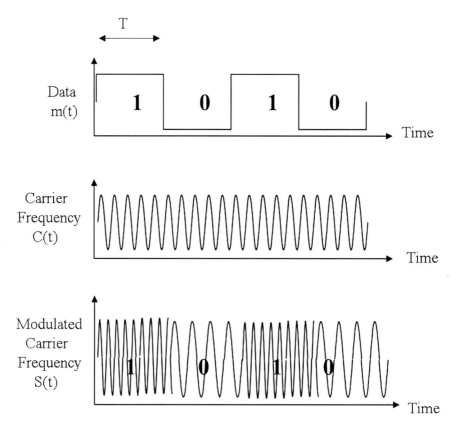

Fig. 5.1 FSK waveforms. The frequency of the carrier changes in accordance with the input digital signal. The amplitude of the carrier remains the same

5.2 Frequency Shift Keying (FSK) Modulation

Frequency shift keying (FSK) is a method of digital modulation that utilizes frequency shifting of the relative frequency content of the signal [1–3]. The signal to be modulated and transmitted is binary, which is encoded before modulation. This is an indispensable task in digital communications, where redundant bits are added with the raw data that enables the receiver to detect and correct bit errors, if they occur during transmission [4–16]. While there are many error-coding schemes available, we will use a simple coding technique, known as "block coding" to illustrate the concept.

Figure 5.2 shows an encoded FSK modulation scheme using (15, 8) block code where an 8-bit data block is formed as M rows and N columns ($M = 4$, $N = 2$). The product $MN = k = 8$ is the dimension of the information bits before coding. Next, a horizontal parity P_H is appended to each row and a vertical parity P_V is appended to each column. The resulting augmented dimension is given by the product $(M + 1)$

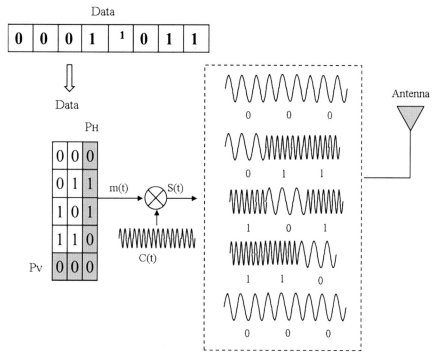

Fig. 5.2 Binary frequency shift keying (BFSK) modulation. The input encoded data block is transmitted row by row. The frequency of the carrier changes in accordance with the input digital signal

$(N + 1) = n = 15$, which is then FSK-modulated and transmitted row by row. The rate of this coding scheme is given by:

$$\text{Code Rate} : r = (MN)/[(M+1)(N+1)] = (4 \times 2)/5 \times 3) = 8/15 \qquad (5.1)$$

The coded bit rate R_{b2} is given by:

$$R_{b2} = \text{Uncoded Bit Rate}/\text{Code Rate} = R_{b1}/r = R_{b1}(15/8) \qquad (5.2)$$

Next, the coded bits are modulated by means of the FSK modulator as shown in the figure. Here,

- The input digital signal is the encoded bit sequence we want to transmit.
- Carrier is the radio frequency without modulation.
- Output is the FSK-modulated carrier, which has two frequencies corresponding to the binary input signal.
- For binary signal 1, the carrier changes to $f_c - \Delta f$.
- For binary signal 0, the carrier changes to $f_c + \Delta f$.
- The total frequency deviation is $2\Delta f$.

As shown in Fig. 3.8, the frequency of the carrier changes in discrete levels, in accordance with the input signals. We have:

- Input Data: $m(t) = 0$ or 1
- Carrier Frequency: $C(t) = A\cos(\omega t)$
- Modulated Carrier: $S(t) = A\cos(\omega - \Delta\omega)t$, For $m(t) = 1$
                      $S(t) = A\cos(\omega + \Delta\omega)t$, For $m(t) = 0$

$$(5.3)$$

where

- $A =$ Frequency of the carrier
- $\omega =$ Nominal frequency of the carrier frequency
- $\Delta\omega =$ Frequency deviation.

$$(5.4)$$

5.3   Frequency Shift Keying (FSK) Demodulation

Once the modulated binary data has been transmitted, it needs to be received and demodulated. This is often accomplished by the use of band-pass filters. In the case of binary FSK, the receiver needs to utilize two band-pass filters that are tuned to the appropriate frequencies. Since the nominal carrier frequency and the frequency deviation are known, this is relatively straightforward. One band-pass filter will be centered at the frequency ω_1 and the other at ω_2. As the signal enters into the receiver, it passes through the respective filter and the corresponding bit value is made. This is shown in Fig. 5.3. In order to assure that the bits are decoded correctly, the frequency deviation needs to be chosen with the limitations of the filters in mind to eliminate crossover.

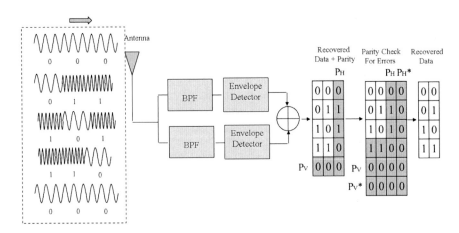

Fig. 5.3 Binary FSK detector utilizing two matched band-pass filters

5.4 FSK Bandwidth

In wireless communications, the scarcity of RF spectrum is well known. For this reason, we have to be vigilant about using transmission bandwidth in error control coding and modulation. The transmission bandwidth depends on the following:

- Spectral response of the encoded data
- Spectral response of the carrier frequency and
- Modulation type.

5.4.1 Spectral Response of the Encoded Data

In digital communications, data is generally referred to as a non-periodic digital signal. It has two values:

- Binary 1 = high, period = T
- Binary 0 = low, period = T

Also, data can be represented in two ways:

- Time-domain representation and
- Frequency-domain representation

The time-domain representation (Fig. 5.4a), known as non-return-to-zero (NRZ), is given by:

$$V(t) = V \quad < 0 < t < T$$
$$= 0 \quad ; \text{elsewhere} \tag{5.5}$$

The frequency-domain representation is given by "Fourier transform":

$$V(\omega) = \int_0^T V \cdot e^{-j\omega t} dt \tag{5.6}$$

$$|V(\omega)| = VT \left[\frac{\sin(\omega T/2)}{\omega T/2} \right] \tag{5.7}$$

$$P(\omega) = \left(\frac{1}{T} \right) |V(\omega)|^2 = V^2 T \left[\frac{\sin(\omega T/2)}{\omega T/2} \right]^2$$

Here, $P(\omega)$ is the power spectral density. This is plotted in Fig. 5.4b. The main lobe corresponds to the fundamental frequency, while side lobes correspond to harmonic components. The bandwidth of the power spectrum is proportional to the

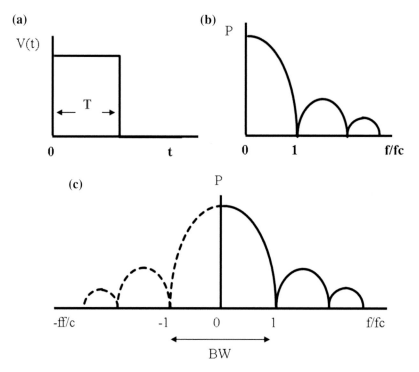

Fig. 5.4 a Discrete-time digital signal, **b** its one-sided power spectral density, and **c** two-sided power spectral density. The bandwidth associated with the non-return-to-zero (NRz) data is $2R_b$, where R_b is the bit rate

frequency. In practice, the side lobes are filtered out, since they are relatively insignificant with respect to the main lobe. Therefore, the one-sided bandwidth is given by the ratio $f/f_b = 1$. In other words, the one-sided bandwidth $= f = f_b$, where $f_b = R_b = 1/T$, T being the bit duration.

The general equation for two-sided response is given by:

$$V(\omega) = \int_{-\infty}^{\infty} V(t) \cdot e^{-j\omega t} dt$$

In this case, $V(\omega)$ is called the two-sided spectrum of $V(t)$. This is due to both positive and negative frequencies used in the integral. The function can be either a voltage or a current. Figure 5.4c shows the two-sided response, where the bandwidth is determined by the main lobe as shown below:

$$\text{Two sided bandwidth (BW)} = 2R_b \, (R_b = \text{Bit rate before coding}) \qquad (5.8)$$

Important Notes:

1. If R_b is the bit rate before coding, and if the data is NRZ, then the bandwidth associated with the raw data will be $2R_b$. For example, if the bit rate before coding is 10 kb/s, then the bandwidth associated with the raw data will be 2×10 kb/s = 20 kHz.
2. If R_b is the bit rate before coding, code rate is r, and if the data is NRZ, then the bit rate after coding will be R_b (coded) = R_b(uncoded)r. The corresponding bandwidth associated with the coded data will be $2R_b$ (coded) = $2R_b$ (uncoded)/r. For example, if the bit rate before coding is 10 kb/s and the code rate $r = 1/2$, the coded bit rate will be R_b (coded) = R_b (uncoded)/r = 10/0.5 = 20 kb/s. The corresponding bandwidth associated with the coded data will be $2 \times 20 = 40$ kHz.

5.4.2 Spectral Response of the Carrier Frequency Before Modulation

A carrier frequency is essentially a sinusoidal waveform, which is periodic and continuous with respect to time. It has one frequency component. For example, the sine wave is described by the following time-domain equation:

$$V(t) = V_p \sin(\omega t_c) \tag{5.9}$$

where

$$V_P = \text{Peak voltage}$$

- $\omega_c = 2\pi f_c$
- f_c = Carrier frequency in Hz

Figure 5.5 shows the characteristics of a sine wave and its spectral response. Since the frequency is constant, its spectral response is located in the horizontal axis and the peak voltage is shown in the vertical axis. The corresponding bandwidth is zero.

5.4.3 FSK Bandwidth at a Glance

In FSK, the frequency of the carrier changes in two discrete levels, in accordance with the input signals. We have:

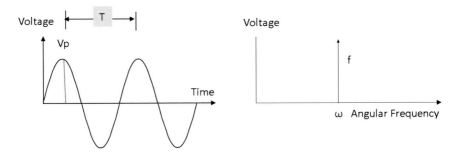

Fig. 5.5 High-frequency carrier response

- Input Data: $m(t) = 0 \text{ or } 1$
- Carrier Frequency: $C(t) = A \cos(\omega t)$
- Modulated Carrier: $S(t) = A \cos(\omega - \Delta\omega)t, \text{For } m(t) = 1$
                        $S(t) = A \cos(\omega + \Delta\omega)t, \text{For } m(t) = 0$

where

- $S(t)$ = The modulated carrier
- A = Amplitude of the carrier
- ω = Nominal frequency of the carrier
- $\Delta\omega$ = Frequency deviation

The spectral response is depicted in Fig. 5.6. Notice that the carrier frequency after FSK modulation varies back and forth from the nominal frequency f_c by $\pm\Delta f_c$, where Δf_c is the frequency deviation. The FSK bandwidth is given by:

$$\begin{aligned} \text{BW} &= 2(f_b + \Delta f_c) \\ &= 2f_b(1 + \Delta f_c/f_b) \qquad\qquad (5.10) \\ &= 2f_b(1 + \beta) \end{aligned}$$

where $\beta = \Delta f/f_b$ is known as the modulation index and f_b is the coded bit frequency (bit rate R_b). The above equation is also known as "Carson's rule."

Problem
Given:

- Bit rate before coding: R_{B1} = 10 kb/s
- Code rate: r = 8/15
- Modulation: FSK
- Modulation index β = 1

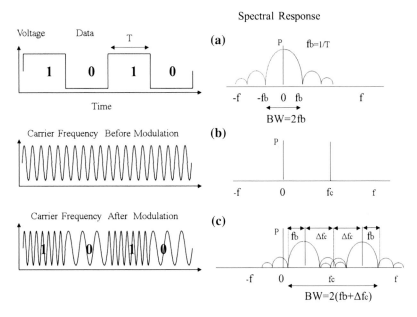

Fig. 5.6 FSK bandwidth at a glance. **a** Spectral response of NRZ data before modulation. **b** Spectral response of the carrier before modulation. **c** Spectral response of the carrier after modulation. The transmission bandwidth is $2(f_b + \Delta f_c)$. f_b where f_b is the bit rate and Δf_c is the frequency deviation $= 1/f_b$ is the bit duration for NRZ data

Find:

(a) the bit rate after coding: R_{B2}
(b) Transmission bandwidth: BW

Solution

(a) Bit rate after coding (R_{B2}):
     $R_{B2} = R_{B1}/r = 10 \text{ kb/s } (15/8) = 18.75 \text{ kb/s}$
(b) Transmission BW $= 2 R_{B2}(1+\beta)$
     $= 2 \times 18.75 \text{ kb/s } (1 + 1) = 75 \text{ kHz}.$

5.5 BER Performance

It is well known that an (n, k) bock code, where $k =$ number of information bits and $n =$ number of coded bits, can correct t errors [4, 5]. A measure of coding gain is then obtained by comparing the uncoded word error WER_U, to the coded word WER_C. We examine this by means of the following analytical means.

Let the uncoded word error be defined as (WER$_U$). Then, with FSK modulation, the uncoded BER will be given by:

$$BER_U = 0.5\,EXP(-E_B/2N_0) \tag{5.11}$$

The probability that the uncoded word (WER$_U$) will be received in error is 1 minus the product of the probabilities that each bit will be received correctly. Thus, we write:

$$WER_U = 1-(1-BER_U)^k \tag{5.12}$$

Let the coded word error be defined as (WER$_C$). Since $n > k$, the ratio of coded bit energy to noise will be modified to E_c/N_0, where $E_c/N_0 = E_b/N_0 + 10\log(k/n)$. Therefore, the coded BER$_c$ will be:

$$BER_C = 0.5\,EXP(-E_c/2N_0) \tag{5.13}$$

The corresponding coded word error rate is:

$$WER_C = \sum_{k=t+1}^{n} \binom{n}{k} BER_C^K (1 - BER_C)^{n-k} \tag{5.14}$$

When BER$_c$ < 0.5, the first term in the summation is the dominant one; therefore, equation can be simplified as

$$WER_C \approx \binom{n}{k} BER_c^k (1 - BER_c)^{n-k} \tag{5.15}$$

Using (15, 8) block code ($n = 15$, $k = 8$, $t = 1$), we obtain the coded and the uncoded WER as shown in Fig. 5.7. Coding gain is the difference in E_b/N_0 between the two curves. Notice that at least 3–4-db coding gain is available in this example where $r = k/n = 8/15$.

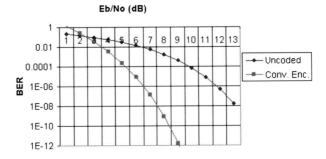

Fig. 5.7 Typical error performance in AWGN

5.6 Conclusions

- This chapter presents FSK modulation and its attributes.
- Numerous illustrations are provided to show how frequency of the carrier changes in discrete levels in accordance with the input digital signal, while the amplitude of the carrier remains the same.
- The Fourier transform is used to derive the spectral components, and FSK bandwidth is calculated.
- Bit error rate (BER) performance is presented.
- These materials have been augmented by diagrams and associated waveforms to make them easier for readers to grasp.

References

1. Smith DR (1985) Digital transmission systems, Van Nostrand Reinhold Co. ISBN: 0442009178
2. Leon W, Couch II (2001) Digital and analog communication systems, 7th edn. Prentice-Hall Inc., Englewood Cliffs. ISBN 0-13-142492-0
3. Sklar B (1988) Digital communications fundamentals and applications. Prentice Hall, Englewood Cliffs
4. Faruque S (2016) Radio frequency channel coding made easy. Springer. ISBN: 978-3-319-21169-5
5. Clark GC et al (1981) Error correction coding for digital communications. Plenum press, New York
6. Ungerboeck G (1982) Channel coding with multilevel/multiphase signals. IEEE Trans Inf Theory IT28:55–67
7. Lin S, Costello DJ Jr (1983) Error control coding: fundamentals and applications. Prentice-Hall, Inc., Englewood Cliffs
8. Blahut RE (1983) Theory and practice of error control codes. Addison-Wesley, Reading
9. van Lint JH (1992) Introduction to coding theory. GTM 86, 2nd edn. Springer, p 31. ISBN 3-540-54894-7
10. Mac Williams FJ, Sloane NJA (1977) The theory of error-correcting codes. North-Holland, p 35. ISBN 0-444-85193-3
11. Huffman W, Pless V (2003) Fundamentals of error-correcting codes. Cambridge University Press. ISBN 978-0-521-78280-7
12. Ebert PM, Tong SY (1968) Convolutional Reed-Soloman codes. Bell Syst Tech 48:729–742
13. Gallager RG (1968) Information theory and reliable communications. Wiley, New York
14. Kohlenbero A, Forney GD Jr (1968) Convolutional coding for channels with memory. IEEE Trans Inf Theory IT-14:618–626
15. Reddy SM (1968) Further results on convolutional codes derived from block codes. Inf Control 13:357–362
16. Reddy SM (1968) A class of linear convolutional codes for compound channels. Technical report, Bell Telephone Laboratories, Holmdel, New Jersey

Chapter 6
Phase Shift Keying (PSK)

Topics

- Introduction
- Binary Phase Shift Keying (BPSK) Modulation
- QPSK Modulation
- 8PSK Modulation
- 16PSK Modulation
- PSK spectrum and Bandwidth
- SSB Spectrum and Bandwidth

6.1 Introduction

In phase shift keying (PSK), the phase of the carrier changes in discrete levels in accordance with the input digital signal, while the amplitude of the carrier remains the same. This is shown in Fig. 6.1, where

- $m(t)$ is the input modulating digital signal,
- $C(t)$ is the carrier frequency, and
- $S(t)$ is the PSK-modulated carrier frequency.

As shown in the figure, the digital binary signal changes the phase of the carrier on two discrete levels. This enables the receiver to extract the digital signal by demodulation. Notice that the phase of the carrier changes in accordance with the input signal, while the amplitude of the carrier does not change after modulation. However, it can be shown that the modulated carrier $S(t)$ contains several spectral components, requiring frequency domain analysis.

In the following sections, the above disciplines in PSK modulation will be presented, along with the respective spectrum and bandwidth. These materials have been augmented by diagrams and associated waveforms to make them easier for readers to grasp.

© The Author(s) 2017   69
S. Faruque, *Radio Frequency Modulation Made Easy*,
SpringerBriefs in Electrical and Computer Engineering,
DOI 10.1007/978-3-319-41202-3_6

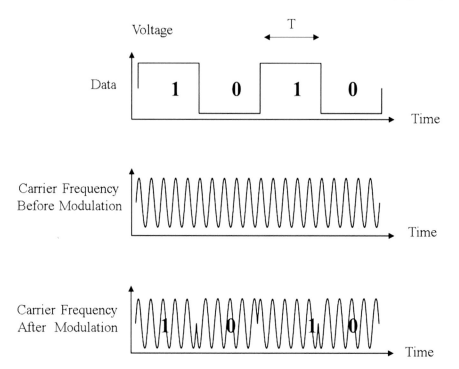

Fig. 6.1 Binary PSK (BPSK) waveforms. The phase of the carrier changes in accordance with the input digital signal. The amplitude of the carrier remains the same

6.2   Binary Phase Shift Keying (BPSK)

6.2.1   BPSK Modulation

Phase shift keying (PSK) is a method of digital modulation that utilizes phase shifting of the relative phase content of the signal [1–3]. The signal to be modulated and transmitted is binary, which is encoded before modulation. This is an indispensable task in digital communications, where redundant bits are added with the raw data that enable the receiver to detect and correct bit errors, if they occur during transmission [4–16]. While there are many error-coding schemes available, we will use a simple coding technique, known as "Block Coding" to illustrate the concept.

Figure 6.2 shows an encoded BPSK modulation scheme using (15, 8) block code where an 8-bit data block is formed as M-rows and N-columns ($M = 4$, $N = 2$). The product $MN = k = 8$ is the dimension of the information bits before coding. Next, a horizontal parity P_H is appended to each row and a vertical parity P_V is appended to each column. The resulting augmented dimension is given by the product $(M + 1)(N + 1) = n = 15$, which is then PSK modulated and transmitted row by row. The rate of this coding scheme is given by

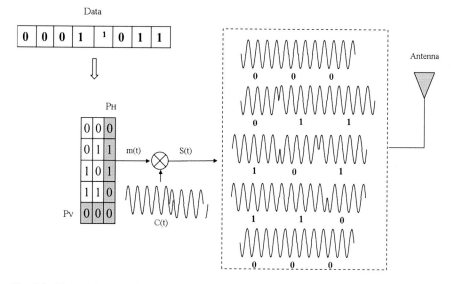

Fig. 6.2 Binary phase shift keying (BPSK) modulation. The input encoded data block is transmitted row by row. The phase of the carrier changes in accordance with the input digital signal

$$\text{Code Rate}: r = (MN)/[(M+1)(N+1)] = (4 \times 2)/5 \times 3) = 8/15 \qquad (6.1)$$

The coded bit rate R_{b2} is given by:

$$R_{b2} = \text{Uncoded Bit Rate/Code Rate} = R_{b1}/r = R_{b1}(15/8) \qquad (6.2)$$

Next, the coded bits are modulated by means of the PSK modulator as shown in the figure. Here,

- The Input digital signal is the encoded bit sequence we want to transmit;
- Carrier is the radio frequency without modulation;
- Output is the PSK-modulated carrier, which has two phases corresponding to the binary input signals;
- For binary signal 0, $\varphi = 0°$; and
- For binary signal 1, $\varphi = 180°$.

As shown in Fig. 6.2, the phase of the carrier changes in two discrete levels, in accordance with the input signals. Here, we have the following:

- Input Data : $m(t) = 0 \text{ or } 1$
- Carrier Frequency : $C(t) = A\cos(\omega_{ct})$
- Modulated Carrier : $S(t) = A_c \cos[\omega_c t + 2\pi/M)m(t)] \quad m(t) = 0, 1, 2, 3, \ldots M-1$

$$(6.3)$$

where,

- A_c = Amplitude of the carrier frequency
- ω_c = Angular frequency of the carrier
- $M = 2, 4, 8, 16, \ldots$
- In BPSK, there are two phases 1 bit/phase$(M = 2)$ (6.4)
- In QPSK, there are four phases, 2-bits/phase, $M = 4$
- In 8PSK, there are 8 phases, 3-bits/phase, $M = 8$
- In 16PSK, there are 16 phases, 4-bits/phase, $M = 16$

We can also represent the BPSK modulator as a signal constellation diagram with $M = 2$, 1 bit per phase. This is shown in Fig. 6.3, where the input raw data, having a bit rate R_{b1}, is encoded by means of a rate r encoder. The encoded data, having a bit rate R_{b2}, = $R_{b1}r$ $(r < 1)$, is modulated by the BPSK modulator as shown in Fig. 6.3.

The BPSK modulator takes one bit at a time to construct the phase constellation having two phases, also known as "Symbols," where each symbol represents one bit. The symbol rate is therefore the same as the encoded bit rate R_{b2}.

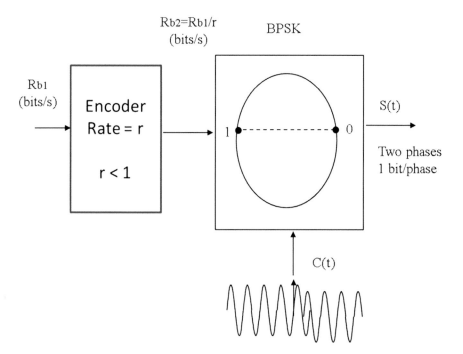

Fig. 6.3 BPSK signal constellation having 2 symbols, 1-bit per symbol

Therefore, the BPSK modulator has the following specifications:

- 2 phases or 2 symbols;
- 1-bit/symbol,

The above specifications govern the transmission bandwidth, as we shall see later.

6.2.2 BPSK Demodulation

Once the modulated binary data has been transmitted, it needs to be received and demodulated. This is often accomplished with the use of a phase detector, typically known as phase-locked loop (PLL). As the signal enters the receiver, it passes through the PLL. The PLL locks the incoming carrier frequency and tracks the variations in frequency and phase. This is known as coherent detection technique, where the knowledge of the carrier frequency and phase must be known to the receiver.

Figure 6.4 shows a simplified diagram of a BPSK demodulator along with the data recovery process. In order to assure that the bits are decoded correctly, the phase deviation needs to be chosen with the limitations of the PLL in mind to eliminate crossover.

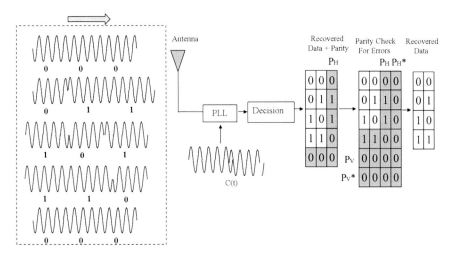

Fig. 6.4 Binary PSK detector showing data recovery process

6.3   QPSK Modulation

In QPSK, the input raw data, having a bit rate R_{b1}, is encoded by a rate r ($r < 1$) encoder. The encoded data, having a bit rate $R_{b2} = R_{b1}/r$, is serial to parallel converted into two parallel streams. The encoded bit rate, now reduced in speed by a factor of two, is modulated by the QPSK modulator as shown in Fig. 6.5.

The QPSK modulator takes one bit from each stream to construct the phase constellation having four phases, also known as "Symbols," where each symbol represents two bits. The symbol rate is therefore reduced by a factor of two. The QPSK modulator has four phases or 4 symbols, 2-bits/symbol as shown in the figure.

Therefore, the QPSK modulator has the following specifications:

- 4 phases or 4 symbols
- 2-bits/symbol

The above specifications govern the transmission bandwidth, as we shall see later.

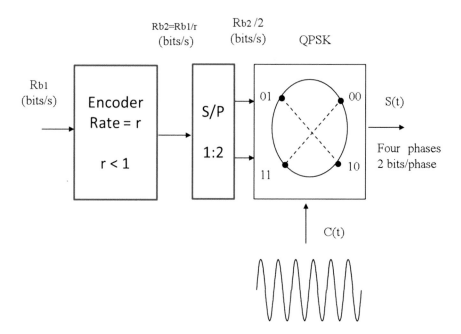

Fig. 6.5 QPSK signal constellation having 4 symbols, 2-bits per symbol

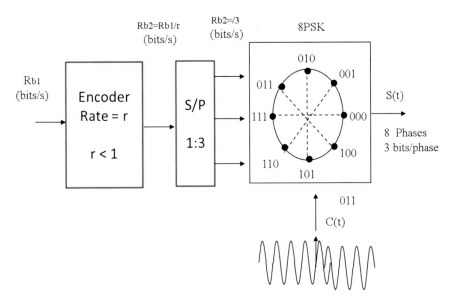

Fig. 6.6 8PSK signal constellation having 8 symbols, 3-bits per symbol

6.4 8PSK Modulation

In 8PSK, the input raw data, having a bit rate R_{b1}, is encoded by a rate r ($r < 1$) encoder. The encoded data, having a bit rate $R_{b2} = R_{b1}/r$, is serial to parallel converted into three parallel streams. The encoded bit rate, now reduced in speed by a factor of three, is modulated by the 8PSK modulator as shown in Fig. 6.6.

The 8PSK modulator takes one bit from each stream to construct the phase constellation having 8 phases, also known as "Symbols," where each symbol represents 3-bits. The symbol rate is therefore reduced by a factor of 3. The 8PSK modulator has 8 phases or 8 symbols, 3-bits/symbol as shown in the figure.

Therefore, the 8PSK modulator has the following specifications:

- 8 phases or 8 symbols;
- 3-bits/symbol.

The above specifications govern the transmission bandwidth, as we shall see later.

6.5 16PSK Modulation

In 16PSK, the input raw data, having a bit rate R_{b1}, is encoded by means of a rate r ($r < 1$) encoder. The encoded data, having a bit rate $R_{b2} = R_{b1}/r$, is serial to parallel converted into four parallel streams. The encoded bit rate, now reduced in speed by a factor of four, is modulated by the 16PSK modulator as shown in Fig. 6.7.

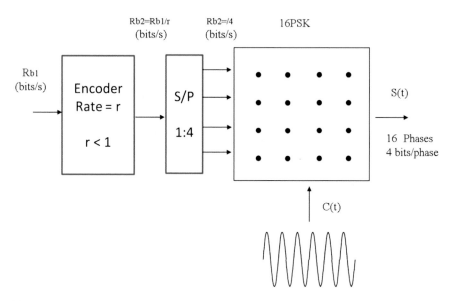

Fig. 6.7 16PSK signal constellation having 16 symbols, 4-bits per symbol. Here, each symbol is represented by a *dot*, where each dot represents 4-bits

The 16PSK modulator takes one bit from each stream to construct the phase constellation having 16 phases, also known as "Symbols," where each symbol represents four bits. The symbol rate is therefore reduced by a factor of four. Therefore, the 16PSK modulator has 16 phases or 16 symbols, 4-bits/symbol as shown in the figure.

Therefore, the 16PSK modulator has the following specifications:

- 16 phases or 16 symbols;
- 4-bits/symbol.

The above specifications govern the transmission bandwidth, as we shall see later.

Table 6.1 shows the number of phases and the corresponding bits per phase for MPSK modulation schemes for $M = 2, 4, 8, 16, 32, 64$, etc.

Table 6.1 MPSK modulation parameters. $M = 2, 4, 8, 16,$ and 32

Modulation	Number of phases φ	Number of bits per phase
BPSK	2	1
QPSK	4	2
8PSK	8	3
16	16	4
32	32	5
64	64	6
:	:	:

6.6 PSK Spectrum and Bandwidth

In wireless communications, the scarcity of RF spectrum is well known. For this reason, we have to be vigilant about using transmission bandwidth in error control coding and modulation. The transmission bandwidth depends on:

- Spectral response of the encoded data;
- Spectral response of the carrier phase; and
- Modulation type.

Let us take a closer look:

6.6.1 Spectral Response of the Encoded Data

In digital communications, data is generally referred to as a non-periodic digital signal. It has two values:

- Binary-1 = High, Period = T
- Binary-0 = Low, Period = T

Also, data can be represented in two ways:

- Time domain representation and
- Frequency domain representation

The time domain representation (Fig. 6.8a), known as non-return-to-zero (NRZ), is given by:

$$V(t) = V \quad <0 < t < T$$
$$= 0 \quad \text{elsewhere} \tag{6.5}$$

The frequency domain representation is given by "Fourier Transform":

$$V(\omega) = \int_{0}^{T} V \cdot e^{-j\omega t} dt \tag{6.6}$$

$$|V(\omega)| = 2\left(\frac{V}{\omega}\right)\sin\left(\frac{\omega T}{2}\right) = VT\left[\frac{\sin(\omega T/2)}{\omega T/2}\right]$$

$$P(\omega) = \left(\frac{1}{T}\right)|V(\omega)|^2 = V^2 T\left[\frac{\sin(\omega T/2)}{\omega T/2}\right]^2 \tag{6.7}$$

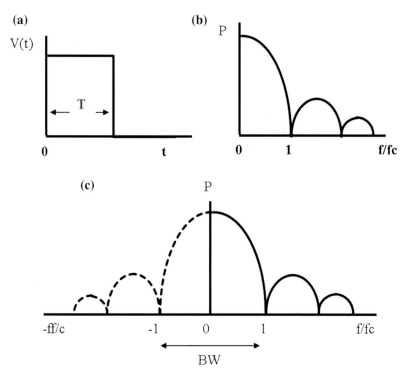

Fig. 6.8 **a** Discrete time digital signal **b** it is one-sided power spectral density and **c** two-sided power spectral density. The bandwidth associated with the non-return-to-zero (NRz) data is $2R_b$, where R_b is the bit rate

Here, $P(\omega)$ is the power spectral density. This is plotted in Fig. 6.8b. The main lobe corresponds to the fundamental frequency, while the side lobes correspond to harmonic components. The bandwidth of the power spectrum is proportional to the frequency. In practice, the side lobes are filtered out, since they are relatively insignificant with respect to the main lobe. Therefore, the one-sided bandwidth is given by the ratio $f/f_b = 1$. In other words, the one-sided bandwidth $= f = f_b$, where $f_b = R_b = 1/T$, T being the bit duration.

The general equation for two-sided response is given by:

$$V(\omega) = \int_{-\infty}^{\infty} V(t) \cdot e^{-j\omega t} dt$$

In this case, $V(\omega)$ is called the two-sided spectrum of $V(t)$. This is due to both positive and negative frequencies used in the integral. The function can be either a voltage or a current. Figure 6.8c shows the two-sided response, where the bandwidth is determined by the main lobe as shown below:

$$\text{Two-sided bandwidth (BW)} = 2R_b \ (R_b = \text{Bit rate before coding}) \qquad (6.8)$$

6.6.2 Spectral Response of the Carrier Before Modulation

A carrier frequency is a sinusoidal waveform, which is periodic and continuous with respect to time. It has one phase component. For example, the sine wave is described by the following time domain equation:

$$V(t) = V_p \sin(\omega_c t) \qquad (6.9)$$

Where

$$Vp = \text{Peak voltage}$$

- $\omega_c = 2\pi f_c$
- $f_c = \text{Carrier phase in Hz}$

Figure 6.9 shows the characteristics of a sine wave and its spectral response. Since the phase is constant, its spectral response is located in the horizontal axis and the peak voltage is shown in the vertical axis. The corresponding bandwidth is zero.

6.6.3 BPSK Spectrum

In BPSK, the input is a digital signal and it contains an infinite number of harmonically related sinusoidal waveforms. This is given by (see Sect. 6.6.1):

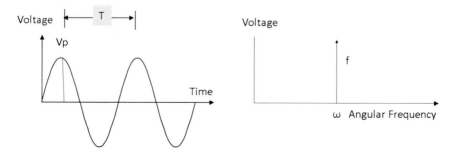

Fig. 6.9 High-frequency carrier response

$$|V(\omega)| = 2\left(\frac{V}{\omega}\right)\sin\left(\frac{\omega T}{2}\right) = VT\left[\frac{\sin(\omega T/2)}{\omega T/2}\right] \qquad (6.10)$$

Here, $V(\omega)$ is the frequency domain representation of the input digital signal, which has a $\sin(x)/x$ response that governs the phase of the carrier frequency.

With $V(t) = m(t)$, we write the following as:

$$S(t) = A_c \cos[\omega_c t + \beta\, m(t)] \qquad (6.11)$$

where β is the phase deviation index of the carrier and $m(t)$ has a $\sin(x)/x$ response, which is given by

$$m(t) = VT\left[\frac{\sin(\omega m T/2)}{\omega m T/2}\right] \qquad (6.12)$$

Therefore, the spectral response after BPSK modulation also has a $\sin(x)/x$ response, which is the shifted version of the NRZ data, centered on the carrier frequency f_c, as shown in Fig. 6.10. The transmission bandwidth associated with the main lobe is given by

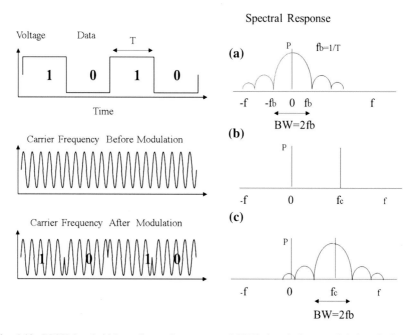

Fig. 6.10 BPSK bandwidth. **a** Spectral response of NRZ data before modulation. **b** Spectral response of the carrier before modulation. **c** Spectral response of the carrier after modulation

$$\text{BW (BPSK)} \approx 2R_{b2}/\text{Bit per Phase}$$
$$\approx 2R_{b2}/1 \approx 2R_{b2} \tag{6.13}$$

where R_{b2} is the coded bit rate (bit frequency). Notice that the BPSK bandwidth is the same the ASK bandwidth. It may be noted that the higher-order spectral components are filtered out.

Problem 6.1

Given:

- Uncoded input bit rate: $R_{b1} = 10$ kb/s
- Code rate: $r = 8/15$
- Carrier frequency $fc = 1$ MHz
- Modulation: BPSK, QPSK, 8PSK, and 16PSK

Find:

(a) Coded bit rate R_{b2}
(b) BPSK bandwidth (BW)
(c) QPSK bandwidth (BW)
(d) 8PSK bandwidth (BW)
(e) 16PSK bandwidth (BW)

Solution:

(a) Coded bit rate: $R_{B2} = R_{B1}/r = 10$ kb/(15/8) = 18.75 kb/s
(b) BPSK bandwidth: $BW = 2R_{B2}/1 = 2 \times 18.75 = 37.5$ kHz
(c) QPSK bandwidth: $BW = 2R_{B2}/2 = 2 \times 18.75/2 = 18.75$ kHz
(d) 8PSK bandwidth: $BW = 2R_{B2}/3 = 2 \times 18.75/3 = 12.5$ kHz
(e) 16PSK bandwidth: $BW = 2R_{B2}/4 = 2 \times 18.755/4 = 9.375$ kHz

NOTE: Higher-order PSK modulation is bandwidth efficient.

Problem 6.2

Given:

- Input bit rate $R_{b1} = 10$ kb/s
- Code rate $r = 1/2$
- Modulation: BPSK, QPSK, 8PSK, 16PSK

Find:

(a) Bit rate after coding R_{b2}
(b) Transmission bandwidth BW

Solution:

(a) $R_{b2} = R_{b1}/r = 10$ kb/s $\times 2 = 20$ kb/s
(b) Transmission bandwidth:

- BPSK $BW = 2R_{b2}$/Bits per Symbol = 2 × 20 kb/s/1 = 40 kHz
- QPSK $BW = 2R_{b2}$/Bits per symbol = 2 × 20 kb/s/2 = 20 kHz
- 8PSK $BW = 2R_{b2}$/Bits per symbol = 2 × 20 kb/s/3 = 13.33 kHz
- 16PSK $BW = 2R_{b2}$/Bits per symbol = 2 × 20 kb/s/4 = 10 kHz

NOTE: Higher-order PSK modulation is bandwidth efficient.

6.7 Conclusions

- This chapter presents PSK modulation and its attributes.
- Numerous illustrations are provided to show how phase of the carrier changes in discrete levels in accordance with the input digital signal, while the amplitude of the carrier remains the same.
- The Fourier transform is used to derive the spectral components and PSK bandwidth is calculated.
- These materials have been augmented by diagrams and associated waveforms to make them easier for readers to grasp.

References

1. Smith DR (1986) Digital transmission systems. Van Nostrand Reinhold Co. ISBN: 0442009178
2. Couch II LW (2001) Digital and analog communication systems, 7th edn. Prentice-Hall, Inc.: Englewood Cliffs. ISBN: 0-13-142492-0
3. Sklar B (1988) Digital communications fundamentals and applications. Prentice Hall
4. Faruque S (2016) Radio frequency channel coding made easy. Springer, ISBN: 978-3-319-21169-5
5. Clark GC et al (1981) Error correction coding for digital communications. Plenum press
6. Ungerboeck G (1982) Channel coding with multilevel/multiphase signals. IEEE Trans Inf Theory IT28:55–67
7. Lin S, Costello DJ Jr (1983) Error control coding: fundamentals and applications. Prentice-Hall, Inc., Englewood Cliffs
8. Blahut RE (1983) Theory and practice of error control codes. reading. Addison-Wesley, Massachusetts
9. van Lint JH (1992) Introduction to coding theory. GTM 86. 2nd ed. Springer. p 31. ISBN 3-540-54894-7
10. Mac Williams FJ, Sloane NJA (1977) The theory of error correcting codes. North-Holland. p 36. ISBN 0-444-85193-3
11. Huffman W, Pless V (2003) Fundamentals of error-correcting codes. Cambridge University Press. ISBN 978-0-521-78280-7
12. Ebert PM, Tong SY (1968) Convolutional Reed-Soloman codes, Bell Syst Tech, pp 729–742
13. Gallager RG (1968) Information theory and reliable communications. John Wiley, New York

14. Kohlenbero A, Forney GD Jr (1968) Convolutional coding for channels with memory, IEEE Trans. Information Theory IT-14, 618–626. 1968
15. M.A. Reddy, S. M, "Further results on convolutional codes derived from block codes. Inf Control 13:357–362
16. Reddy SM (1968) A class of linear convolutional codes for compound channels. Technical Report, Bell Telephone Laboratories, Holmdel, New Jersey

Chapter 7
N-Ary Coded Modulation

Topics

- Introduction
- N-ary Complementary Convolutional Coding and M-ary Modulation
- N-ary Complementary Orthogonal Coding and M-ary Modulation
- Coding Gain and Bandwidth
- Conclusions

7.1 Introduction

N-ary coded modulation is a multi-level channel coding and multi-level modulation technique, where instead of coding one bit at a time two or more bits are encoded simultaneously and then modulated by means of M-ary modulation. In its most basic construction, the input serial data is converted into several parallel streams. These parallel bit streams, now reduced in speed, are mapped into a bank of N unique convolutional or orthogonal codes. The coded information bits are then modulated by an M-ary PSK modulator and transmitted through a channel. At the receiver, the decoder recovers the data by means of code correlation. A lookup table at the receiver contains the input/output bit sequences. Upon receiving an encoded data pattern, the receiver validates the received data pattern by means of code correlation.

In this chapter, we will present the key concept, underlying principles and practical application of N-ary coded modulation schemes, offering spectrum efficiency with improved error correction capabilities. Construction of N-ary coded modulation schemes based on convolutional and orthogonal codes will be presented to illustrate the concept.

© The Author(s) 2017
S. Faruque, *Radio Frequency Modulation Made Easy*,
SpringerBriefs in Electrical and Computer Engineering,
DOI 10.1007/978-3-319-41202-3_7

7.2   N-Ary Convolutional Coding and M-Ary Modulation

7.2.1   Background

In convolutional coding, a sequence of data signals is transformed into a longer sequence that contains enough redundancy to protect the data [1–6]. This type of error control is also classified as forward error control coding (FECC), because these methods are often used to correct errors that are caused by channel noise. In a typical convolutional encoder, k information bits enter into the encoder sequentially. The convolutional encoder generates n parity bits as encoded bits ($n > k$). The code rate is defined as $r = k/n$. The coded information bits are modulated and transmitted through a channel.

Another class of convolutional codes, known as parallel concatenated codes or turbo codes [7–9], is popular in wireless communications because of its superior error control capabilities. Yet, turbo codes are limited to two convolutional codes connected in parallel, where delays are introduced due to concatenation and interleaving, thereby limiting high-speed data communications in wireless communications. In addition, it is also too complex to realize.

In this section, a modified technique, defined as N-ary complementary convolutional coding, along with M-ary modulation, is presented to overcome these problems. N-ary complementary convolutional coding is a multi-level convolutional code, where more than two convolutional codes can be used simultaneously to further enhance coding gain without concatenation and interleaving.

7.2.2   Generation of Complementary Convolutional Codes

In a typical convolutional encoder, k information bits enter into the encoder sequentially. The encoder generates n parity bits as encoded bits ($n > k$). The code rate is defined as $r = k/n$. The proposed N-ary convolutional codes are a block of n-convolutional codes and their complements.

In its most basic construction, as shown in Fig. 7.1, a typical convolutional encoder is taken as the basis to generate a block of n-convolutional codes. Next, each convolutional code is inverted to construct a block of n-antipodal codes. Therefore, a block of N-ary convolutional code has n-convolutional codes and n-antipodal codes, for a total of $2n = N$ complementary convolutional codes. In short, we shall define this as N-ary complementary convolutional codes.

Let us consider Fig. 7.1 once again and assume that the input bit sequence to the encoder is:

$$m(t) = 1\ 0\ 1 \qquad\qquad (7.1)$$

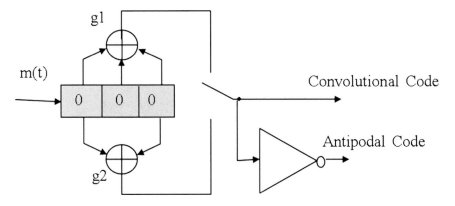

Fig. 7.1 Generation of N-ary complimentary convolutional codes

Which is described by the following polynomial?

$$m(X) = 1 + 0X + 1X^2$$
$$= 1 + X^2 \tag{7.2}$$

The encoder is described by the following generator polynomials:

$$g1(X) = 1 + X + X^2$$
$$g2(X) = 1 + X^2 \tag{7.3}$$

Then, the product of polynomials for the encoder can be described as follows:

$$m(X)g1(X) = (1 + X^2)(1 + X + X^2) = 1 + X + X^3 + X^4$$
$$m(X)g2(X) = (1 + X^2)(1 + X^2) = 1 + X^4 \tag{7.4}$$

with $X^2 + X^2 = 0$, the output bit sequence can be found as $U(X) = m(X)g1(X)$ multiplexed with $m(X)g2(X)$, where $m(X)$ is the input bit sequence [7]. We write the above two equations as follows:

$$
\begin{array}{llllllllll}
m(X)g1(X) &=& 1 &+& 1X &+& 0X^2 &+& 1X^3 &+& 1X^4 \\
m(X)g2(X) &=& 1 &+& 0X &+& 0X^2 &+& 0X^3 &+& 1X^4
\end{array}
$$

-- ------------

$$U1U2 = (1,1) + (1,0) + (0,0) + (1,0) + (1,1)$$

$$(7.5)$$

Convolutional Code   Antipodal Code

Convolutional Code	Antipodal Code
00 00 00 00 00	11 11 11 11 11
11 10 11 00 00	00 01 00 11 11
00 11 10 11 00	11 00 01 00 11
11 01 01 11 00	00 10 10 00 11
00 00 11 10 11	11 11 00 0100
11 10 00 10 11	00 01 11 01 00
00 11 01 01 11	11 00 10 10 00
11 01 10 01 11	00 10 01 10 00

Fig. 7.2 N-ary complimentary convolutional code blocks

Taking only the coefficients, we obtain the desired convolutional code sequence corresponding to the input bit sequence $m(T) = 101$ as follows:

$$\text{Convolutional Code} = 1\,1 \quad 1\,0 \quad 0\,0 \quad 1\,0 \quad 1\,1 \qquad (7.6)$$

The corresponding antipodal code is obtained simply by inverting the convolutional code as shown in the figure. This is given by:

$$\text{Antipodal Code} = 0\,0 \quad 0\,1 \quad 1\,1 \quad 0\,1 \quad 0\,0 \qquad (7.7)$$

As can be seen, each 3-bit data corresponds to a unique 10-bit convolutional code. Antipodal code is just the inverse of the convolutional codes. Since a 3-bit sequence has 8 combinations, the above convolutional encoder generates 8 unique convolutional codes and 8 unique antipodal codes, for a total of 16 complementary convolutional codes. These code blocks are displayed in Fig. 7.2.

7.2.3 2-Ary Convolutional Coding with QPSK Modulation

A 2-ary ($N = 2$) coded QPSK modulator can be constructed by inverse multiplexing the incoming traffic into 6-parallel streams as shown in Fig. 7.3. These bit streams, now reduced in speed by a factor of 6, are partitioned into two 8×3 data blocks. The first 8×3 data block maps the 8×10 convolutional code block, and the next 8×3 data block maps the next 8×10 antipodal code block. These code blocks are stored in two 8×10 ROMs. The output of each ROM is a unique 10-bit

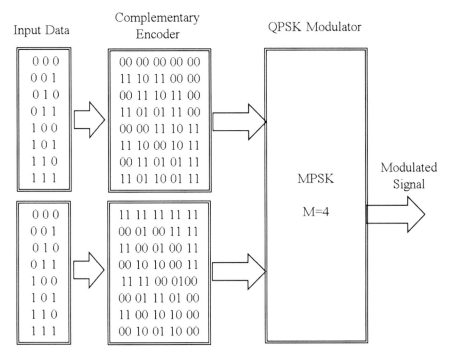

Fig. 7.3 2-ary convolutional coding with QPSK modulation

convolutional/antipodal code, which is modulated by a QPSK modulator using the same carrier frequency. The code rate is given by:

$$\text{Code Rate:} \, r = 6/10 = 3/5 \tag{7.8}$$

Since there are two convolutional waveforms (one convolutional and one antipodal), the number of errors that can be corrected is doubled. Moreover, the transmission bandwidth is also reduced, which is given by:

Transmission bandwidth: $BW = 2R_{b2}$/bits per symbol Hz

$$B = 2R_{b2}/2 = R_{b2} \tag{7.9}$$

where $R_{b2} = R_{b1}/r$. R_{b2} is the coded bit rate, and R_{b1} is the uncoded bit rate.

7.2.4 4-Ary Convolutional Coding with 16PSK Modulation

A 4-ary ($N = 4$) coded 16PSK modulator can be constructed by inverse multiplexing the incoming traffic into 8-parallel streams as shown in Fig. 7.4. These bit streams, now reduced in speed by a factor of 8, are partitioned into four 4×2 data

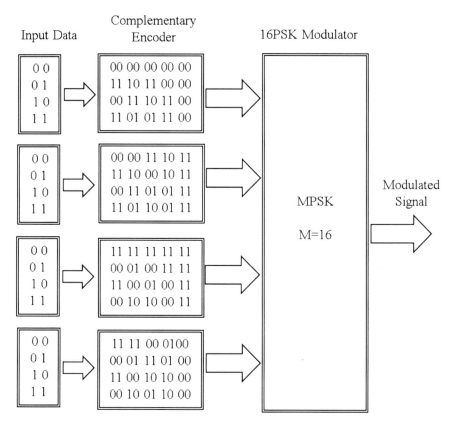

Fig. 7.4 4-ary convolutional coding with 16PSK modulation

blocks. Each 4×2 data block maps corresponding 4×10 convolutional code blocks. These code blocks are stored in four 4×10 ROMs. The output of each ROM is a unique 10-bit convolutional/antipodal code, which is modulated by a 16PSK modulator using the same carrier frequency. The code rate is given by:

$$\text{Code Rate:} r = 8/10 = 4/5 \tag{7.10}$$

Since there are four convolutional waveforms (two convolutional and two antipodal), the number of errors that can be corrected is quadrupled. Moreover, the transmission bandwidth is further reduced, which is given by:

Transmission bandwidth: $BW = 2R_{b2}/$bits per symbol Hz

$$= 2R_{b2}/4 = R_{b2}/2 \tag{7.11}$$

where $R_{b2} = R_{b1}/r$. R_{b2} is the coded bit rate, and R_{b1} is the uncoded bit rate.

7.3 N-Ary Convolutional Decoder

7.3.1 Correlation Receiver

Decoding is a process of code correlation. In this process, the receiver compares the received data with the expected data set to recover the actual data. The expected data is stored into a lookup table [see Table 7.1]:

- The lookup table at the receiver contains the input/output bit sequences.
- For $m = 3$, there are 8 possible output combinations of 3-bit data.
- For each combination of 3-bit data, there is a unique encoded 10-bit data (see table).
- The receiver receives one of 8 output sequences.
- Upon receiving an encoded data pattern, the receiver validates the received data pattern by means of code correlation.

Let us examine the correlation process using the following example:

- The input bit pattern $m = 0\ 1\ 1$
- Encoded transmit data: $U = 00\ 11\ 01\ 01\ 11$
- Received data with errors $U^* = 00\ 11\ 01\ 01\ \mathbf{00}$

Notice that the last two bits are in error, identified in bold. Now, let us determine how the receiver recovers the correct data, where the actual input data is $m = 0\ 1\ 1$. This is a correlation process, requiring several tests to validate the actual data. The correlation process is described below:

Test 0:

This test compares the received data with the 1st row of data stored in the lookup table and counts the number of positions it does not match. This is accomplished by MOD2 operation (EXOR operation). The result is presented below:

- Received data: 00 11 01 01 **00**
- 1st row of data in the lookup table: <u>00 00 00 00 00</u>
- Mod-2 Add: 00 11 01 01 00

Table 7.1 Lookup table a

Input (m)	Output (U)
0. 0 0 0	00 00 00 00 00
1. 0 0 1	00 00 11 10 11
2. 0 1 0	00 11 10 11 00
3. 0 1 1	00 11 01 01 11
4. 1 0 0	11 10 11 00 00
5. 1 0 1	11 10 00 10 11
6. 1 1 0	11 01 01 11 00
7. 1 1 1	11 01 10 01 11

- Correlation value = 4 (count the number of 1's in MOD2 Add)
- Verdict: No match, continue search.

Test-0: Look Up Table

Input (m)	Output (U)	Correlation Value
0. 0 0 0	00 00 00 00 00	4
1. 0 0 1	00 00 11 10 11	
2. 0 1 0	00 11 10 11 00	
3. 0 1 1	00 11 01 01 11	
4. 1 0 0	11 10 11 00 00	
5. 1 0 1	11 10 00 10 11	
6. 1 1 0	11 01 01 11 00	
7. 1 1 1	11 01 10 01 11	

Received Data: 00 11 01 01 00

Test 1:

This test compares the received data with the 2nd row of data stored in the lookup table and counts the number of positions it does not match. This is accomplished by MOD2 operation (EXOR operation). The result is presented below:

- Received data: 00 11 01 01 00
- 2nd row of data in the lookup table: <u>00 00 11 10 11</u>
- Mod-2 Add: 00 11 10 111 11
- Correlation value = 7
- Verdict: No match, continue search

Test 1: Look Up Table

Input (m)	Output (U)	Correlation Value
0. 0 0 0	00 00 00 00 00	
1. 0 0 1	00 00 11 10 11	7
2. 0 1 0	00 11 10 11 00	
3. 0 1 1	00 11 01 01 11	
4. 1 0 0	11 10 11 00 00	
5. 1 0 1	11 10 00 10 11	
6. 1 1 0	11 01 01 11 00	
7. 1 1 1	11 01 10 01 11	

Received Data: 00 11 01 01 00

The remaining tests may be conducted by using the same method to reach the final verdict as given below:

Final Verdict:

Collect the correlation values and validate the data that indicates the lowest correlation value. This is presented in the following lookup table.

Final Verdict: Look Up Table

Input (m)	Output (U)	Correlation Value
0. 0 0 0	00 00 00 00 00	4
1. 0 0 1	00 00 11 10 11	7
2. 0 1 0	00 11 10 11 00	3
3. 0 1 1	00 11 01 01 11	2
4. 1 0 0	11 10 11 00 00	5
5. 1 0 1	11 10 00 10 11	8
6. 1 1 0	11 01 01 11 00	4
7. 1 1 1	11 01 10 01 11	7

Accept ⟶ (row 3)   ⟵ Accept

In examining the above table, we find that:

- The lowest correlation value is 2.
- The corresponding data is $m = 0\ 1\ 1$.
- This is the data which has been transmitted to the receiver.

7.3.2 Error Correction Capabilities of N-Ary Convolutional Codes

An n-bit convolutional code is generated by a rate ½ encoder of constraint length K where

$$n = 4K - 2 \qquad (7.12)$$

In the above equation, K is the length of the shift register and the factor -2 is due to the initial content of the shift register, which is 000. For example, a rate ½ convolutional encoder with $K = 3$ generates 00 when the initial content of the shift register is 000. Therefore, the value of n is reduced by 2. Therefore, each 3-bit data sequence corresponds to a unique 10-bit convolutional code. The antipodal code is just the inverse of the convolutional codes. Since a 3-bit sequence has 8 combinations, the above convolutional encoder generates 8 unique convolutional codes and 8 unique antipodal codes, for a total of 16 complementary convolutional codes.

These code blocks are displayed in Fig. 7.2. As can be seen, each code is unique and there is a minimum distance d_{min} between any two convolutional codes where

$$d_{min} \leq n/2 \qquad (7.13)$$

This distance property can be used to detect an impaired received code by setting a threshold midway between two convolutional codes. It can be shown that an n-bit convolutional code can correct the errors, where

Table 7.2 Error correction capabilities of N-ary convolutional codes

Constraint length (K)	Number of errors corrected if the modulation scheme is (BPSK)	Number of errors corrected if the modulation scheme is (QPSK)	Number of errors corrected if the modulation scheme is (16PSK)
3	2	4	8
5	4	8	16
7	6	12	24
9	8	16	32

$$t \leq d_{\min}/2 \leq n/4 \leq (4K - 2)/4 \tag{7.14}$$

where

- t = Number of errors that can be corrected by means of a single convolutional code
- n = Code length
- K = Constraint length

The proposed N-ary coded modulation is a multi-level channel coding technique, where instead of coding one bit at a time two or more bits are encoded simultaneously. Therefore, we modify the above equation to obtain

$$N t \leq N d_{\min}/2 \leq Nn/4 \leq N(4K - 2)/4 \tag{7.15}$$

Table 7.2 displays error correction capabilities of N-ary convolutional codes.

7.4 N-Ary Orthogonal Coding and M-Ary Modulation

7.4.1 Background

Orthogonal codes are used in CDMA cellular communications for spectrum spreading and user ID [10–13]. The use of orthogonal codes for forward error control coding has also been investigated by a limited number of authors [14–16]. This chapter presents a method of channel coding based on N-ary orthogonal codes [17–20]. N-ary orthogonal coding is a multi-level channel coding technique, where instead of coding one bit at a time two or more bits are encoded simultaneously. This type of channel coding is bandwidth efficient and further enhances the coding gain with bandwidth efficiency. In the proposed scheme, the input serial data is converted into several parallel streams. These parallel bit streams, now reduced in speed, are mapped into N-ary orthogonal codes. The coded information bits are then modulated by means of MPSK modulator and transmitted through a channel. At the receiver, the decoder recovers the data by means of code correlation. The proposed coded modulation scheme is bandwidth efficient and offers a coding gain near the Shannon's limit.

Orthogonal Code   Antipodal Code

```
0 0 0 0 0 0 0 0              1 1 1 1 1 1 1 1
0 1 0 1 0 1 0 1              1 0 1 0 1 0 1 0
0 0 1 1 0 0 1 1              1 1 0 0 1 1 0 0
0 1 1 0 0 1 1 0              1 0 0 1 1 0 0 1
0 0 0 0 1 1 1 1              1 1 1 1 0 0 0 0
0 1 0 1 1 0 1 0              1 0 1 0 0 1 0 1
0 0 1 1 1 1 0 0              1 1 0 0 0 0 1 1
0 1 1 0 1 0 0 1              1 0 0 1 0 1 1 0
```

Fig. 7.5 Bi-orthogonal code set for $n = 8$. An 8-bit orthogonal code has 8 orthogonal codes and 8 antipodal codes for a total of 16 bi-orthogonal codes

7.4.2   Orthogonal Codes

Orthogonal codes are binary-valued and have equal numbers of 1's and 0's. Antipodal codes, on the other hand, are just the inverse of orthogonal codes. Antipodal codes are also orthogonal among them. Therefore, an n-bit orthogonal code has n-orthogonal codes and n-antipodal codes, for a total of 2n bi-orthogonal codes. For example, an 8-bit orthogonal code has 8 orthogonal codes and 8 antipodal codes, for a total of 16 bi-orthogonal codes, as shown in Fig. 7.5 [12]. Similarly, a 16-bit orthogonal code has 16 orthogonal codes and 16 antipodal codes for a total of 32 bi-orthogonal codes, as shown in Fig. 7.6.

We note that orthogonal codes are essentially (n, k) block codes, where a k-bit data set is represented by a unique n-bit orthogonal code ($k < n$). We now show that code rates such as rate ½, rate ¾, and rate 1 are indeed available out of orthogonal codes. The principle is presented below.

7.4.3   2-Ary Orthogonal Coding with QPSK Modulation

A 2-ary orthogonal coded QPSK modulator with $n = 8$, having 16 complementary orthogonal codes, can be constructed by inverse multiplexing the incoming traffic into 6-parallel streams, as shown in Fig. 7.7. These bit streams, now reduced in speed by a factor of 6, are partitioned into two data blocks. Each data block is mapped into an 8×8 code block, as depicted in the figure.

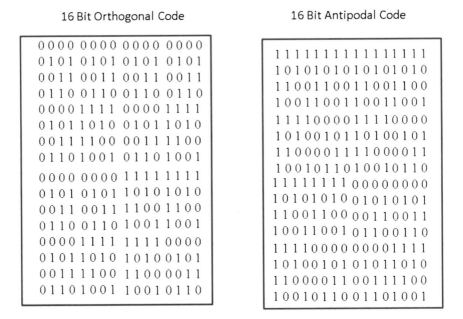

16 Bit Orthogonal Code

16 Bit Antipodal Code

```
0000 0000 0000 0000
0101 0101 0101 0101
0011 0011 0011 0011
0110 0110 0110 0110
0000 1111 0000 1111
0101 1010 0101 1010
0011 1100 0011 1100
0110 1001 0110 1001
0000 0000 1111 1111
0101 0101 1010 1010
0011 0011 1100 1100
0110 0110 1001 1001
0000 1111 1111 0000
0101 1010 1010 0101
0011 1100 1100 0011
0110 1001 1001 0110
```

```
1111 1111 1111 1111
1010 1010 1010 1010
1100 1100 1100 1100
1001 1001 1001 1001
1111 0000 1111 0000
1010 0101 1010 0101
1100 0011 1100 0011
1001 0110 1001 0110
1111 1111 0000 0000
1010 1010 0101 0101
1100 1100 0011 0011
1001 1001 0110 0110
1111 0000 0000 1111
1010 0101 0101 1010
1100 0011 0011 1100
1001 0110 0110 1001
```

Fig. 7.6 Bi-orthogonal code set for $n = 16$. A 16-bit orthogonal code has 16 orthogonal codes and 16 antipodal codes for a total of 32 bi-orthogonal codes

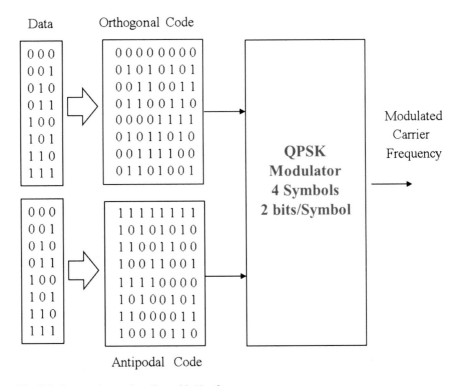

Data   Orthogonal Code

```
000          0 0 0 0 0 0 0 0
001          0 1 0 1 0 1 0 1
010          0 0 1 1 0 0 1 1
011          0 1 1 0 0 1 1 0
100          0 0 0 0 1 1 1 1
101          0 1 0 1 1 0 1 0
110          0 0 1 1 1 1 0 0
111          0 1 1 0 1 0 0 1
```

QPSK
Modulator
4 Symbols
2 bits/Symbol

Modulated
Carrier
Frequency

```
000          1 1 1 1 1 1 1 1
001          1 0 1 0 1 0 1 0
010          1 1 0 0 1 1 0 0
011          1 0 0 1 1 0 0 1
100          1 1 1 1 0 0 0 0
101          1 0 1 0 0 1 0 1
110          1 1 0 0 0 0 1 1
111          1 0 0 1 0 1 1 0
```

Antipodal Code

Fig. 7.7 2-ary orthogonal coding with $N = 2$

According to the communication scheme, when a 3-bit data pattern needs to be transmitted, the corresponding orthogonal/antipodal code is transmitted instead, requiring a QPSK modulator. The modulated waveforms are in orthogonal space and have fewer errors. The code rate is given by:

$$\text{Code Rate: } r = 6/8 = 3/4$$

Since there are two orthogonal waveforms (one orthogonal and one antipodal), the number of errors that can be corrected is doubled. Moreover, the transmission bandwidth is also reduced, which is given by:

Transmission Bandwidth: $BW = 2R_{b2}$/bits per symbol Hz

$$= 2R_{b2}/2 = R_{b2} \text{ Hz} \qquad (7.16)$$

where $R_{b2} = R_{b1}/r$. R_{b2} is the coded bit rate and R_{b1} is the uncoded bit rate.

7.4.4 4-Ary Orthogonal Coding with 16PSK Modulation

A 4-ary orthogonal coded 16PSK modulator, with $n = 8$, having 16 complementary orthogonal codes, can be constructed by inverse multiplexing the incoming traffic into 8-parallel streams, as shown in Fig. 7.8. The bit streams, now reduced in speed by a factor of 8, are partitioned into four data blocks. Each data block is mapped into a 4×8 code block as depicted in the figure. According to the communication scheme, when a 2-bit data pattern needs to be transmitted, the corresponding orthogonal/antipodal code is transmitted instead, requiring a 16PSK modulator. The modulated waveforms are in orthogonal space and have fewer errors. The code rate is given by:

$$\text{Code Rate: } r = 8/8 = 1 \qquad (7.17)$$

This is achieved without bandwidth expansion.

Since there are four orthogonal waveforms (two orthogonal and two antipodal), the number of errors that can be corrected is quadrupled. Moreover, the transmission bandwidth is further reduced, which is given by:

Transmission bandwidth: $BW = 2R_{b2}$/bits per symbol Hz

$$= 2R_{b2}/4 = R_{b2}/2 \text{ Hz} \qquad (7.18)$$

where $R_{b2} = R_{b1}/r$. R_{b2} is the coded bit rate, and R_{b1} is the uncoded bit rate.

7.4.5 2-Ary Orthogonal Decoding

Decoding is a correlation process. Notice that the entire biorthogonal code block is partitioned into two code blocks. Each code block represents a data block as shown in Fig. 7.9. Upon receiving an impaired code, the receiver compares it with each

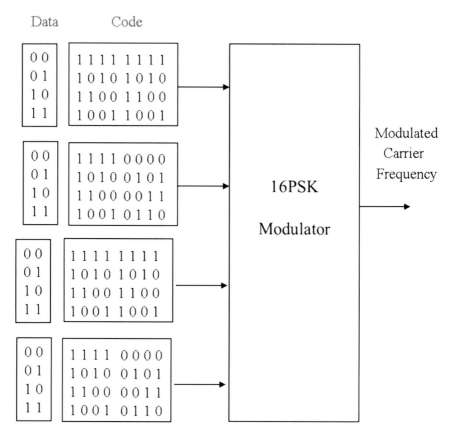

Fig. 7.8 N-ary orthogonal coding with $N = 4$

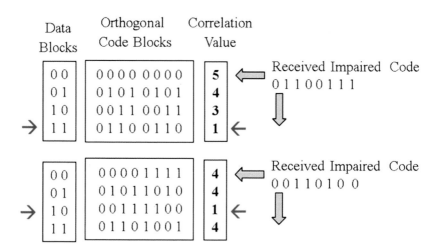

Fig. 7.9 2-ary orthogonal decoding

entry in the code block and appends a correlation value for each comparison. A valid code is declared when the closest approximation is achieved. As can be seen, the minimum correlation value in each block is 1 as depicted in the figure. The corresponding data is 1 1, 1 0, 1 1, 1 0, respectively.

Since there are two orthogonal waveforms (one orthogonal and one antipodal), the number of errors that can be corrected is given by $2 \times 1 = 2$. Moreover, the bandwidth is also reduced, as presented below:

$$BW = 2Rb2/\text{bits per symbol} = 2_{Rb2}/2 = R_{b2} \qquad (7.19)$$

7.4.6   4-Ary Orthogonal Decoding

Decoding is a correlation process. Notice that the entire biorthogonal code block is partitioned into four code blocks. Each code block represents a data block, as shown in Fig. 7.10. Upon receiving an impaired code, the receiver compares it with each entry in the code block and appends a correlation value for each comparison. A valid code is declared when the closest approximation is achieved. As can be seen, the minimum correlation value in each block is 1 as depicted in the figure. The corresponding data is 1 1, 1 0, 1 1, 1 0, respectively.

Since there are four orthogonal waveforms (two orthogonal and two antipodal), the number of errors that can be corrected is given by $4 \times 1 = 4$. Moreover, the bandwidth is further reduced as presented below:

$$BW = 2Rb2/\text{bits per symbol} = 2_{Rb2}/4 = R_{b2}/2 \qquad (7.20)$$

7.4.7   Error Correction Capabilities of N-Ary Orthogonal Codes

Error correction capabilities of orthogonal codes have been discussed earlier [16]. We present it again for convenience. An n-bit orthogonal code has $n/2$ 1's and $n/2$ 0's; i.e., there are $n/2$ positions where 1's and 0's differ. Therefore, the distance between two orthogonal codes is $d = n/2$. This distance property can be used to detect an impaired received code by setting a threshold midway between two orthogonal codes as shown in Fig. 7.11, where the received code is shown as a dotted line. This is given by:

$$d_{th} = \frac{n}{4} \qquad (7.21)$$

where n is the code length and d_{th} is the threshold, which is midway between two valid orthogonal codes. Therefore, for the given 8-bit orthogonal code, we have

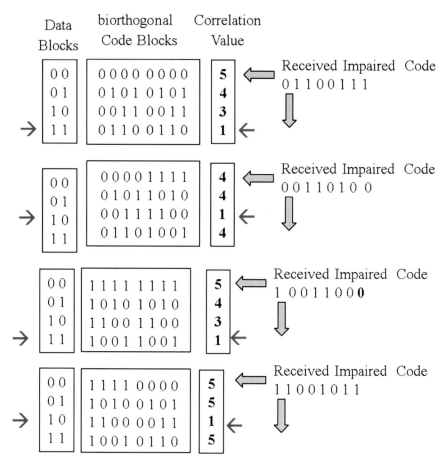

Data Blocks	biorthogonal Code Blocks	Correlation Value	
0 0	0 0 0 0 0 0 0 0	5	← Received Impaired Code
0 1	0 1 0 1 0 1 0 1	4	0 1 1 0 0 1 1 1
1 0	0 0 1 1 0 0 1 1	3	
1 1	0 1 1 0 0 1 1 0	1	

0 0	0 0 0 0 1 1 1 1	4	← Received Impaired Code
0 1	0 1 0 1 1 0 1 0	4	0 0 1 1 0 1 0 0
1 0	0 0 1 1 1 1 0 0	1	
1 1	0 1 1 0 1 0 0 1	4	

0 0	1 1 1 1 1 1 1 1	5	← Received Impaired Code
0 1	1 0 1 0 1 0 1 0	4	1 0 0 1 1 0 0 0
1 0	1 1 0 0 1 1 0 0	3	
1 1	1 0 0 1 1 0 0 1	1	

0 0	1 1 1 1 0 0 0 0	5	← Received Impaired Code
0 1	1 0 1 0 0 1 0 1	5	1 1 0 0 1 0 1 1
1 0	1 1 0 0 0 0 1 1	1	
1 1	1 0 0 1 0 1 1 0	5	

Fig. 7.10   4-ary orthogonal decoding

Fig. 7.11 Distance
properties of orthogonal codes

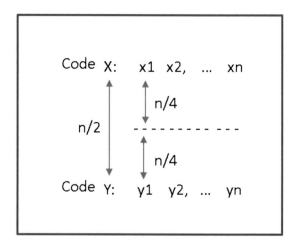

$d_{th} = 8/4 = 2$. This mechanism offers a decision process, where the incoming impaired orthogonal code is examined for correlation with the neighboring codes for a possible match.

The received code is examined for correlation with the neighboring codes for a possible match. The acceptance criterion for a valid code is that an n-bit comparison must yield a good autocorrelation value; otherwise, a false detection will occur. The following correlation process governs this, where an impaired orthogonal code *is* compared with a pair of n-bit orthogonal codes to yield:

$$R(x, y) = \sum_{i=1}^{n} x_i y_i \geq (n - d_{th}) + 1 \qquad (7.22)$$

where x and y are two n-bit orthogonal codes, $R(x, y)$ is the autocorrelation function, n is the code length, and d_{th} is the threshold as defined earlier. Since the threshold is in the midway between two valid codes, an additional 1-bit offset is added to Eq. 7.21 for reliable detection. The average number of errors that can be corrected by means of this process can be estimated by combining Eq. 7.21 and Eq. 7.22, yielding:

$$t = n - R(x, y) = \frac{n}{4} - 1 \qquad (7.23)$$

In the above equation, t is the number of errors that can be corrected by means of an n-bit orthogonal code. For example, a single-error-correcting orthogonal code can be constructed by means of an 8-bit orthogonal code ($n = 8$). Similarly, a three-error-correcting orthogonal code can be constructed by means of a 16-bit orthogonal code ($n = 16$) and so on. Table 7.3 shows a few orthogonal codes and the corresponding error-correcting capabilities.

Problem 7.1 This problem relates to N-ary convolutional codes and M-ary PSK modulation.

   Given:

- Input bit rate before coding: $R_{b1} = 10$ kb/s
- $K = 3$, Rate ½ complementary convolutional codes
- 4-ary convolutional coding ($N = 4$)

Table 7.3 Error correction capabilities of N-ary orthogonal codes

Code length n	Number of errors corrected per code t	Number of errors corrected by N-ary code Nt $N = 1, 2, 4, 8, \ldots$
8	1	N
16	3	3 N
32	7	7 N
64	15	15 N

Find:

(a) Modulation level M
(b) Code rate r
(c) Bit rate after coding R_{b2}
(d) Transmission bandwidth BW
(e) Number of errors corrected

Solution:

(a) A 4-ary ($N = 4$) coded modulator can be constructed by inverse multiplexing the incoming traffic into 8-parallel stream. These bit streams, now reduced in speed by a factor of 8, are partitioned into four 4×2 data blocks. Each 4×2 data block maps a corresponding 4×10 convolutional code blocks. These code blocks are stored in four 4×10 ROMs. The output of each ROM is a unique 10-bit convolutional/antipodal code, which is modulated by a $2^4 = 16$ PSK modulator using the same carrier frequency. Therefore, $M = 16$.
(b) The code rate is given by:

$$\text{Code Rate:} \, r = 8/10 = 4/5$$

(c) Bit rate after coding $R_{b2} = R_{b1}/r = (5/4) \, 10 \text{ kb/s} = 12.5 \text{ kb/s}$
(d) Transmission bandwidth: $BW = 2R_{b2}/\text{bits per symbol Hz}$

$$= 2R_{b2}/4 = R_{b2}/2 = (12.50 \, \text{kb/s})/2 = 6.25 \, \text{kHz}$$

(e) Since there are four convolutional waveforms (two convolutional and two antipodal), the number of errors that can be corrected is quadruple, $4t = 4 \times 2 = 8$ [Each code corrects 2 errors.]

Problem 7.2 This example relates to N-ary orthogonal codes and M-ary PSK modulation.

Given:

• Input bit rate before coding: $R_{b1} = 10$ kb/s
• $n = 8$ orthogonal codes having 16 complementary orthogonal codes
• 4-ary orthogonal Coding

Find:

(a) Modulation level M
(b) Code rate r
(c) Bit rate after coding R_{b2}
(d) Transmission bandwidth BW
(e) Number of errors corrected

Solution:

(a) A 4-ary orthogonal coded modulator, with $n = 8$, having 16 complementary orthogonal codes, can be constructed by inverse multiplexing the incoming traffic into 8-parallel streams. The bit streams, now reduced in speed by a factor of 8, are partitioned into four data blocks. Each data block is mapped into a 4×8 code block. According to the communication scheme, when a 2-bit data pattern needs to be transmitted, the corresponding orthogonal/antipodal code is transmitted instead, requiring a $2^4 = 16$ PSK modulator. Therefore, $M = 416$.

(b) The code rate is given by:

$$\text{Code Rate: } r = 8/8 = 1$$

(c) The bit rate after coding is given by:

$$R_{b2} = R_{b1}r = 10\,\text{kb/s} \ (r = 1)$$

(d) The transmission bandwidth is given by:

$$BW = 2Rb2/\text{bits per symbol Hz}$$
$$= 2R_{b2}/4 = 2 \times 10\,\text{kb/s}/4 = 5\,\text{k Hz}$$

(e) Since there are four orthogonal waveforms (two orthogonal and two antipodal), the number of errors that can be corrected is quadrupled, i.e.,

$$4t = 4 \times 1 = 4$$

7.5   Conclusions

A method of coded modulation, based on N-ary complementary codes and M-ary PSK (MPSK) modulation, is presented. N-ary coded modulation is a multi-level channel coding and multi-level modulation technique, where instead of coding one bit at a time two or more bits are encoded simultaneously and then modulated by means of M-ary modulation. At the receiver, the decoder recovers the data by means of code correlation. A lookup table at the receiver contains the input/output bit sequences. Upon receiving an encoded data pattern, the receiver validates the received data pattern by means of code correlation. Construction of N-ary coded modulation schemes based on convolutional and orthogonal codes is presented to illustrate the concept. The proposed coded modulation schemes are bandwidth efficient and offer a coding gain near the Shannon's limit.

References

1. Ebert PM, Tong SY (1968) Convolutional Reed-Soloman codes, Bell Syst Tech, pp 729–742
2. Gallager RG (1968) Information theory and reliable communications. Wiley, New York
3. Kohlenbero A,.Forney GD Jr (1968) Convolutional coding for channels with memory, IEEE Trans Inf Theory IT-14:618–626
4. Reddy SM (1968) Further results on convolutional codes derived from block codes. Inf Control 13:357–362
5. Reddy SM (1968) A class of linear convolutional codes for compound channels. Technical Report, Bell Telephone Laboratories, Holmdel, New Jersey
6. Reddy SM, Robinson JP (1968) A construction for convolutional codes using block codes. Inf Control 12:55–70
7. Berrou C, Glavieux A, Thitimajshima P (2010) Near Shannon limit error—correcting (PDF). Retrieved 11 February 2010
8. Berrou C (2010) The ten-year-old turbo codes are entering into service, Bretagne, France. Retrieved11 February 2010
9. McEliece RJ, MacKay DJC, Cheng J-F (1998) Turbo decoding as an instance of Pearl's "belief propagation" algorithm. IEEE J Sel Areas Commun 16(2):140–152, doi:10.1109/49. 661103, ISSN 0733-8716
10. IS-95 Mobile station—base station compatibility standard for dual mode wide band spread
11. Spectrum Cellular Systems, TR 55, PN-3115, March 15 (1993)
12. Samuel CY (1998) CDMA RF system engineering. Artech House Inc
13. Vijay KG (1999) IS-95 CDMA and cdma 2000. Prentice Hall
14. Bernard S (1988) Digital communications fundamentals and applications. Prentice Hall
15. Faruque S, Faruque S, Semke W (2013) Orthogonal on-off keying for free-space laser communications with ambient light cancellation. SPIE Optical Eng J 52(9), September 26
16. Faruque S (2016) Radio frequency channel coding made easy, Springer, ISBN: 978-3-319-21169-5
17. Faruque S (1999) Battlefield wideband transceivers based on combined N-ary orthogonal signaling and M-ary PSK modulation, SPIE proceedings, vol. 3709 digitization of the battle space 1 V, pp 123–128
18. Faruque S et al (2001) Broadband wireless access based on code division parallel access, US Patent No. 6208615, March 27
19. Faruque S et al (2001) Bi-orthogonal code division multiple access system, US Patent No. 6198719, March 6
20. Faruque S (2003) Code division multiple access cable modem, US Patent No. 6647059, November